EL-121

FUNDAMENTALS OF DIGITAL ELECTRONICS

[2 Credits]

Electronic Science : Paper-I

For

First Year B.Sc. Semester-II

As Per New Syllabus of CBCS Pattern

From June - 2019

Prof. (Dr.) P. B. Buchade
M.Sc., M.Phil., Ph.D.
Principal and Head
Dept. of Electronic Science,
Abasaheb Garware College,
Karve Road, Pune - 411004

Prof. (Dr.) M. L. Dongare
M.Sc., D.H.E., Ph.D.
Head, Dept. of Electronic Science,
Rayat Shikshan Sanstha's,
S. M. Joshi College,
Hadapsar, Pune - 411028

Dr. J. A. Bangali
M. Sc., M. Phil, Ph.D.
Dept. of Electronic Science,
Kaveri College of Science and Commerce,
Pune – 411038

S. R. Chaudhari
M. Sc., M.Phil.
Vice Principal and Head,
Head, Dept. of Electronic Science,
Modern College, Shivajinagar,
Pune – 411005

N5030

F.Y.B.Sc : Fundamentals of Digital Electronics

First Edition : **November 2019**   **ISBN 978-93-89686-32-6**

© : **Authors**

Published By:
NIRALI PRAKASHAN
Abhyudaya Pragati, 1312, Shivaji Nagar
Off J.M. Road, PUNE – 411005
Tel - (020) 25512336/37/39, Fax - (020) 25511379
Email : niralipune@pragationline.com

➤ DISTRIBUTION CENTRES

PUNE

Nirali Prakashan : 119, Budhwar Peth, Jogeshwari Mandir Lane, Pune 411002,
(For orders within Pune) Maharashtra, Tel : (020) 2445 2044, Mobile : 9657703145
Email : niralilocal@pragationline.com

Nirali Prakashan : S. No. 28/27, Dhayari, Near Asian College Pune 411041
(For orders outside Pune) Tel : (020) 24690204; Mobile : 9657703143
Email : bookorder@pragationline.com

MUMBAI

Nirali Prakashan : 385, S.V.P. Road, Rasdhara Co-op. Hsg. Society Ltd.,
Girgaum, Mumbai 400004, Maharashtra;
Mobile : 9320129587 Tel : (022) 2385 6339 / 2386 9976,
Fax : (022) 2386 9976
Email : niralimumbai@pragationline.com

➤ DISTRIBUTION BRANCHES

JALGAON

Nirali Prakashan : 34, V. V. Golani Market, Navi Peth, Jalgaon 425001,
Maharashtra, Tel : (0257) 222 0395, Mob : 94234 91860;
Email : niralijalgaon@pragationline.com

KOLHAPUR

Nirali Prakashan : New Mahadvar Road, Kedar Plaza, 1st Floor Opp. IDBI Bank,
Kolhapur 416 012, Maharashtra. Mob : 9850046155;
Email : niralikolhapur@pragationline.com

NAGPUR

Nirali Prakashan : Above Maratha Mandir, Shop No. 3, First Floor,
Rani Jhanshi Square, Sitabuldi, Nagpur 440012, Maharashtra
Tel : (0712) 254 7129;
Email : niralinagpur@pragationline.com

DELHI

Nirali Prakashan : 4593/15, Basement, Agarwal Lane, Ansari Road, Daryaganj
Near Times of India Building, New Delhi 110002
Mob : 08505972553, Email : niralidelhi@pragationline.com

BENGALURU

Nirali Prakashan : Maitri Ground Floor, Jaya Apartments, No. 99, 6th Cross,
6th Main, Malleswaram, Bengaluru 560003, Karnataka;
Mob : 9449043034
Email: niralibangalore@pragationline.com
Other Branches : Hyderabad, Chennai

niralipune@pragationline.com | www.pragationline.com
Also find us on 🅕 **www.facebook.com/niralibooks**

Preface ...

We have great pleasure to present this book on **'Fundamentals of digital Electreonics'** which is written as per the revised syllabus of Savitribai Phule Pune University for F.Y.B.Sc. Electronics. The authors have sincerely put their efforts to give complete information of the subject as prescribed in the syllabus in a simple manner.

Electronics have made tremendous revolution in last decade. The majority of this revolution is in the digital world. Students entering in the field of Electronics should have understanding of basic fundamentals of digital electronics.

This text book has been prepared keeping in mind the need of subject and syllabus specified by SPPU.

The First Chapter describes basics of digital electronics which includes number system, logic gates and Boolean algebra.

Second chapter gives details of combinational logic circuits, its construction, operating principal and applications.

The third chapter gives details of sequential circuits. The fundamental block such as Flip-flops, its different types, counters, shift registers are explained in detail in this chapter.

After studying all this, student will get an idea and understanding of fundamentals of digital electronics. To give something extra, to think and to explore, the topic 'Think Over It' element is added in this book.

We are thankful to the publishers Shri Dineshbhai Furia and Shri Jignesh Furia and the staff of Nirali Prakashan specially Mr. Santosh Bare, Mrs. Manasi Pingle, Ms. Chaitali Takle, Mr. Ravindra Walodare for the great efforts they have taken to publish this book in time.

 We hope this book will be very helpful to the students of Electronic Science.

All valuable suggestions from the readers of this book are always welcome.

AUTHORS

Syllabus ...

Unit 1 : BASICS OF DIGITAL ELECTRONICS   (16 L)

Number Systems : Decimal, Binary, Hexadecimal, BCD, Gray code and their inter-conversions. ASCII, Complements (1's, 2's), Rules of binary addition, subtraction.

Logic Gates : Positive and negative logic, AND, OR, NOT, EX-OR, NAND, NOR, EX-NOR and truth tables, NAND and NOR universal gates.

Boolean Algebra and Theorems : Boolean theorems, De-Morgan's laws. Digital logic gates, Multi level NAND and NOR gates. Standard representation of logic functions (SOP and POS), Minimization techniques (Karnaugh map method : 3 variables), Do not care condition.

Basic concepts of Arithmetic and Logical Unit (ALU).

Unit 2 : COMBINATIONAL LOGIC CIRCUITS   (10 L)

Adder - half and full adder, Subtractor - half and full subtractors, Parallel binary adder, Magnitude comparator, Digital lock using magnitude comparator, Multiplexers (2 : 1, 4 : 1) and demultiplexers (1 : 2, 4 : 1), Encoder (8-line-to-3-line) and Decoder (3-line-to-8-line). Parity generator and checker.

Unit 3 : SEQUENTIAL LOGIC CIRCUITS   (10 L)

Flip-flops and truth tables : S-R FF, J-K FF, T and D type FFs, Master-slave FFs, Flip-flop as memory device.

Shift registers and their types, Serial to parallel and parallel to serial converters using shift registers. Counters : Asynchronous - Mod 16, Mod-10, Mod-8, Up-down couner.

Synchronous - Ring counter, Event counter.

•••

Contents ...

•••

Chapter **1**...

Basics of Digital Electronics

The most commonly used system of numerals is the Hindu – Arabic numeral system. Two Indian mathematicians are credited with developing it. Aryabhata, developed the place-value notation in the 5th century, and a century later Brahmagupta introduced the symbol for zero.

The modern binary number system, the basis for binary code, was invented by Gottfried Leibniz in 1689.

George Boole (2 November 1815 – 8 December 1864) was a English mathematician, philosopher and logician, most of whose short career was spent as the first professor of mathematics at Queen's College, Cork in Ireland. He worked in the fields of differential equations and algebraic logic, and is best known as the author of The Laws of Thought (1854) which contains Boolean algebra.

1.1 Introduction to Digital Electronics

- Electronic circuits are classified into two broad categories; Digital and Analog.

- Digital electronics involves quantities with discrete values and analog electronics involves quantities with continuous values.

- The word digital comes from the source word digit and digitus (the Latin word for finger), as fingers are used for discrete counting.

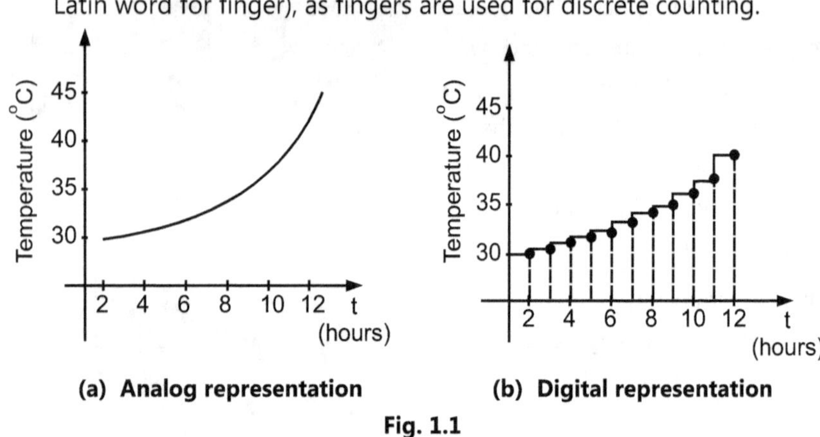

(a) **Analog representation**   (b) **Digital representation**

Fig. 1.1

- An analog quantity is one having continuous value. A digital quantity is one having a discrete set of values.

- Most of the physical quantities appearing in nature are analog. For example, temperature, pressure of atmosphere.

- If we consider a temperature of a given day, it does not go from 40°C to 41°C, rather it takes all infinite values in between.

- However, instead of recording temperature on a continuous basis, if we take temperature reading after every hour we will get discrete values of the temperature.

- In Fig. 1.1 (a) the temperature is recorded continuously and as observed from it the temperature changes smoothly from 35°C to 45°C.

- In Fig. 1.1 (b) the temperature is recorded after an hour and it shows discrete values.

Advantages of Digital Data Over Analog Data :

- It is found that digital data can be processed and transmitted more efficiently and reliably than analog data.

- Digital data can be stored more compactly and reproduced with greater accuracy and clarity than when it is in analog form.

- Important reason for the widespread use of digital techniques and systems are :
  1. The basic building block used in digital circuits operate in one of the two states, known as ON and OFF which results in a very simple operation.
  2. There are only a few basic operations in digital circuits which are very easy to understand.
  3. The logic involved in digital techniques is based on Boolean algebra which is quite simple to understand.
  4. The large number of ready to use electronic circuits in the form of single chip known as IC's are available for performing various operations. They are highly reliable, accurate, small in size and the speed of operation is very high.
  5. Digital circuits are highly suitable for computers, calculators, telephone etc.
  6. The display of digital data is very convenient, and elegant.
- Digital Systems/Electronics contains circuits and systems in which there are only two possible states.
- These states are represented by two different voltage levels, usually referred to as High level and Low level.
- The two states can also be represented by switch with open and closed state or with lamp in the ON and OFF state. Such two state systems are called a ***binary system***.
- The binary system uses symbols 1 and 0 called as ***binary bits***, which is a contraction of the word's binary digit.
- In digital circuit, two different voltage levels are used to represent the two bits. Normally, 1 is represented by the higher voltage which is referred as high and a 0 is represented by the lower voltage level which is referred as low.
- The voltages used to represent a 1 and a 0 are called logic levels. In general, a voltage range is specified or assigned to represent logic levels rather than assigning a single voltage. Fig. 1.2 shows digital signal representation.
- As shown in Fig. 1.2, in a practical digital circuit a High can be 'any voltage between a specified minimum value and a specified maximum value', similarly a Low can be 'any voltage between a specified minimum and a specified maximum value'.

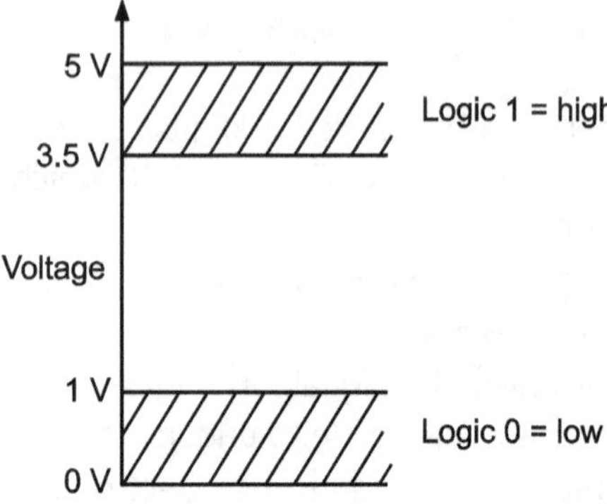

Fig. 1.2 : Voltage ranges for logic 1 and 0

1.2 Introduction to Number Systems

- After hearing the word number we immediately think of the most familiar decimal number system with its 10 digits 0, 1, 2, 3, 4, 5, 6, 7, 8 and 9. However, it is only customary and not necessary to use these symbols.
- Basically, a *number system* is 'an ordered set of symbols known as digits', and the total number of symbols used in a system defines the type of number system. Some of the commonly used number systems are binary, octal and hexadecimal number system.
- Each number system has predefined rules for performing arithmetic operations like addition, subtraction, multiplication etc.
- In this chapter, we will study different number systems, their interconversion and the different codes used in the digital system.

1.3 Decimal Number System

- In decimal number system, an ordered set of ten symbols 0, 1, 2, 3, 4, 5, 6, 7, 8 and 9 are used to specify the quantities. The symbols used are known as digits. Since deci-stands for ten, decimal number system uses ten basic symbols. The number of different basic symbols used in a system is known as *base* or *radix* of a number system.
- Since decimal number system has 10 basic symbols, the base or radix of this number system is 10. The radix of binary system is 2, octal system is 8 and that of hexadecimal is 16.
- Table 1.1 shows the representation of different quantities in decimal number system, where black dots are used to represent the quantities. We can use decimal digit 0 through 9 to represent quantities upto 9. In order to represent 10 quantities, the decimal 10 is obtained by combining the second digit followed by the first.

Table 1.1 : Decimal digits

Quantity	Symbol
None	0
•	1
••	2
•••	3
••••	4
•••••	5
••••••	6
•••••••	7
••••••••	8
•••••••••	9
••••••••••	10
•••••••••••	11

- Since decimal number system uses 10 symbols, we can express decimal number (integer or a whole number) in units, tens, hundreds and so on.

- For example, the decimal number 298 decomposes into,

$$298 = 200 + 90 + 8$$

  In powers of 10, this becomes

$$298 = 2 \times 10^2 + 9 \times 10^1 + 8 \times 10^0$$

- Thus, the decimal number can be expressed as sum of powers of ten and in above example, the digit 2 has weight 200, the digit 9 has weight 90 and the digit 8 has weight 8.

- As we go from right to the left, the weight also known as place value of decimal digit increases in terms of powers of 10. However, the decimal number may have an integer and a fractional part. The integer and the fractional part are separated by a point known as the *decimal point* e.g. decimal number 52.38, where 52 is an integer part and 38 is the fractional part of the given number.

- The weights assigned in case of mixed number are

$$\text{etc.} \leftarrow 10^3 \; 10^2 \; 10^1 \; 10^0 \cdot 10^{-1} \; 10^{-2} \; 10^{-3} \rightarrow \text{etc.}$$

Decimal point

Hence, the number 52.38 can be expressed as,

$$52.38 = 5 \times 10^1 + 2 \times 10^0 + 3 \times 10^{-1} + 8 \times 10^{-2}$$

$$= 50 + 2 + \frac{3}{10} + \frac{8}{100} = 50 + 2 + 0.3 + 0.08$$

$$= 52.38$$

- There are some other number systems that are also used to represent quantities. Some of the commonly used number systems are binary, octal and hexadecimal number system. These systems are widely used in computers, logic circuits, microprocessors etc.

1.4 Binary Number System

- Decimal number system is not suitable for numerical operation and computation using electronic machine like computers, microprocessors etc.

- It is very much essential to convert decimal numbers to binary numbers that will solve this problem. The main advantage of using binary number system is that, it minimizes the number of lines in a computer system.

- The binary number system is a code that uses only two basic symbols i.e. 0 and 1. Generally, in digital electronics, '0' represents the low state and '1' the high state. These symbols are known as *bits*.

- These states 1 and 0 are also referred as ON and OFF, HIGH or LOW, magnetized and demagnetized, white or black etc.

- We can form many other number systems by selecting a different group of basic symbols or digits.

- The **base or radix** of number system refers to the number of basic symbols used. For instance, ten is the base of the decimal number system because this system uses ten digits 0 through 9. The binary number system has a base of two because the only two digits 0 and 1 are used.

- All binary numbers consist of a string of 0s and 1s. Examples are 10, 101, 1001 and which are read as one zero, one zero one, one zero zero one, to avoid confusion with decimal numbers.

- An easy way to remember how to write binary sequence is as follows. If we begin counting decimal numbers starting from 0_{10} and only the count is taken into account having only 0 and 1 then we get, 0, 1, ... 10, 11, ... 100, 101 ... 110, 111 ... 1000, 1001, ... 1010, 1011, ... 1100, 1101, ... 1110, 1111.

The above numbers are tabulated in systematic form in the following Table 1.2.

Table 1.2 : Binary number system

Decimal number	Binary number
0	0
1	1
2	10
3	11
4	100
5	101
6	110
7	111
8	1000
9	1001
10	1010
11	1011
12	1100
13	1101
14	1110
15	1111

- As we see from the table, only two symbols are available. The group of four bits is known as **nibble** and group of eight bits is known as **byte.**
- e.g. 1001, 1110 etc. are nibbles and 10101011, such group is called as byte.
- A binary number can be represented as,

$$N = d_n \times 2^n + d_{n-1} \times 2^{n-1} +$$

$$\ldots d_1 \times 2^1 + d_0 \times 2^0 + d_{-1} 2^{-1} + \ldots + d_{-m} 2^{-m}$$

where d_n, d_{n-1} ... $d_{-(m-1)}$, d_{-m} are binary symbols (0 or 1) and n, m are integers.

- Like the decimal system binary system is also positionally weighted. However, the position value of each bit corresponds to some power of 2.

- In each binary number, the value increases by power of 2 starting with 0 to left of binary point and decreases to the right of the binary point starting with power of -1.

- The left most bit is known as **Most Significant Bit (MSB)** and the right most bit is known as **Least Significant Bit (LSB).** MSB has highest weight while LSB has lowest weight. The position value (or weight) of each bit along with 8 bit 1010.1001 is shown below :

MSB   LSB

1	0	1	0	.	1	0	0	1
2^3	2^2	2^1	2^0	$\downarrow$	2^{-1}	2^{-2}	2^{-3}	2^{-4}

Binary
point

- As seen from the above example, fourth bit to the left of binary point carries the maximum weight (i.e. has the highest value) and is called Most Significant Bit (MSB) or most significant digit.

- Similarly, the fourth bit to the right of the binary point is called Least Significant Bit (LSB) or least significant digit.

- The position value of different bits are given by ascending power of 2 to the left of binary point and descending power of 2 to the right of the binary point. The different digit position in general form may be stated as given below :

$$\leftarrow \quad 2^4 \quad 2^3 \quad 2^2 \quad 2^1 \quad 2^0 \quad \cdot \quad 2^{-1} \quad 2^{-2} \quad 2^{-3} \quad 2^{-4} \quad \rightarrow$$

$\downarrow$
Binary
point

$$\leftarrow \quad 16 \quad 8 \quad 4 \quad 2 \quad 1 \quad\quad 1/2 \quad 1/4 \quad 1/8 \quad 1/16 \quad \rightarrow$$

Position Value:

- If 2 bits are used, the highest count is upto 3. If 3 bits are used it is upto 7, and for 4 bit highest count is 15, means in general we can say,

- Highest decimal count = $2^n - 1$, where n is number of bits.

$\quad\quad$ If n = 5, i.e. with 5 bits we can count from 0 through 31.

$$2^5 - 1 = 32 - 1$$
$$= 31 \text{ is highest count}$$

- If we prepare table for 5 bit binary number presentation, we will have the following Table 1.3.

Table 1.3

Decimal number	Binary number
0	00000
1	00001
2	00010
3	00011
4	00100
5	00101
6	00110
7	00111
8	01000
9	01001
...	...
...	...
...	...
...	...
...	...
...	...
30	11110
31	11111

An easy way to remember how to write a binary sequence such as in Table 1.3, for five bit example we have,

1. The LSB i.e. right most column binary number begins with 0 and alternates each bit.

2. The second LSB i.e. next column have two 0s and alternates every two bits.

3. The third LSB have four 0s and alternates every four bits.

4. The fourth LSB have eight 0s and alternates every eight bits.

5. The next column i.e. MSB have sixteen 0s and alternates every sixteen bits.

1.5 Binary to Decimal Conversion

- Any binary number can be converted into its equivalent decimal number using the weights assigned to each bit position. The rightmost bit (LSB) in a binary number has weight of $2^0 = 1$. The weights increase by power of two for each bit from right to left.

  e.g. Binary number 111 becomes,

$$\text{Binary number} \quad \rightarrow \quad 1 \ \ 1 \ \ 1$$
$$\text{Binary weight} \quad \rightarrow \quad 2^2 \ 2^1 \ 2^0$$
$$= \ 4 + 2 + 1$$
$$111 \ = \ 7$$

Example 1 : Convert the binary number 1111 to decimal.

Solution :

$$\begin{array}{ll}
\text{Binary number} & 1 \ \ 1 \ \ 1 \ \ 1 \\
\text{Binary weight} & 2^3 \ 2^2 \ 2^1 \ 2^0 \\
\text{Weight value} & 8 \ \ 4 \ \ 2 \ \ 1
\end{array}$$

$$= \ 1 \times 8 + 1 \times 4 + 1 \times 2 + 1 \times 1$$
$$= \ 8 + 4 + 2 + 1$$
$$= \ (15)_D \ \text{ or } \ (15)_{10}$$

It can be represented as

$$(1111)_B \ = \ (15)_D$$
$$\text{or} \qquad (1111)_2 \ = \ (15)_{10}$$

The subscript D or 10 identifies decimal number and subscript B or 2 identifies binary number.

Example 2 : Convert the binary number 110101 to decimal number.

Solution :

$$\begin{array}{ll}
\text{Binary number} & 1 \ \ 1 \ \ 0 \ \ 1 \ \ 0 \ \ 1 \\
\text{Binary weight} & 2^5 \ 2^4 \ 2^3 \ 2^2 \ 2^1 \ 2^0 \\
\text{Weight value} & 32 \ 16 \ 8 \ \ 4 \ \ 2 \ \ 1
\end{array}$$

$$= \ 1 \times 32 + 1 \times 16 + 0 + 1 \times 4 + 0 + 1 \times 1$$
$$= \ 32 + 16 + 4 + 1 \ = \ (53)_{10}$$

We can also use streamlined method by the following procedure,

$$\begin{array}{ll}
\text{Step 1} & 1 \ \ 1 \ \ 0 \ \ 1 \ \ 0 \ \ 1 \\
\text{Step 2} & 32 \ 16 \ 8 \ \ 4 \ \ 2 \ \ 1 \\
\text{Step 3} & 32 \ 16 \ \cancel{8} \ \ 4 \ \ \cancel{2} \ \ 1
\end{array}$$

In step 3 if zero appears in digit position, cross out the decimal weight for that position.

$$\text{Step 4} \quad 32 + 16 + 4 + 1 = (53)_{10}$$

$$(110101)_2 = (53)_{10}$$

$$\text{or} \quad (110101)_B = (53)_D$$

The binary numbers we have seen so far have been whole numbers. Fractional number can also be represented in a binary by placing bits to the right of binary point.

In general, binary number with fractions can be written as

$$2^n \ldots 2^4\, 2^3\, 2^2\, 2^1\, 2^0 \cdot 2^{-1}\, 2^{-2}\, 2^{-3} \ldots 2^{-n}$$

$$\downarrow$$

binary point

This indicates that all the bits to the left of binary point have weights that are positive power of two.

All bits to the right of binary point have weights that are negative power at two or fractional weight.

$$\cdot \quad 2^{-1},\ 2^{-2},\ 2^{-3},\ 2^{-4}\ \text{etc.}$$

$$\downarrow$$

binary point

$$\cdot \quad \frac{1}{2},\ \frac{1}{4},\ \frac{1}{8},\ \frac{1}{16}\ \text{etc.}$$

$$\cdot \quad 0.5,\ 0.25,\ 0.125,\ 0.0625\ \text{etc.}$$

$$\downarrow$$

binary point

Example 3 : Determine the decimal number represented by binary fraction 0.101.

Solution : First determine the weight of each bit and then sum the weight times the bit.

Binary weight	2^{-1}	2^{-2}	2^{-3}
Weight value	0.5	0.25	0.125
Binary number	0.1	0	1

$$= 1 \times 0.5 + 0 \times 0.25 + 1 \times 0.125$$

$$= 0.5 + 0 + 0.125$$

$$= 0.625$$

Example 4 : Determine the decimal value of the binary number 0.1011.

Solution :

Binary weight	2^{-1}	2^{-2}	2^{-3}	2^{-4}
Weight value	0.5	0.25	0.125	0.0625
Binary value	0.1	0	1	1

$$= 1 \times 0.5 + 0 \times 0.25 + 1 \times 0.125 + 1 \times 0.0625$$

$$= 0.5 + 0 + 0.125 + 0.0625$$

$$= 0.5 + 0.125 + 0.0625$$

$$= (0.6875)_{10}$$

Mixed numbers (numbers with integer and fractional part) convert each part according to the rules developed.

$$1100 \cdot 1011 = 1 \times 8 + 1 \times 4 + 0 + 0 + 1 \times 0.5 + 0$$
$$+ 1 \times 0.125 + 1 \times 0.0625$$
$$= 8 + 4 + 0.5 + 0.125 + 0.0625 = 12.6875$$
$$(1100 \cdot 1011)_2 = (12.6875)_{10}$$

Example 5 : Perform the following : $(11010.010)_2 = (?)_{10}$.

Solution : 11010.010 $\leftarrow$ given number

Step :

(i)   Write the weight 16 8 4 2 1 0.5 0.125 0.0625

(ii) Discard the weight whose coefficients are zero

16 8 ~~4~~ 2 1 ~~0.5~~ 0.125 ~~0.625~~

(iii) Add the remaining weight 16 + 8 + 2 + 0.125 = 26.125

$\therefore$ $(11010.010)_2 = (26.125)_{10}$

1.6 Decimal to Binary Conversion

- Any decimal number can be converted into its equivalent binary number. One way to convert a decimal number into binary equivalent is reverse process of binary to decimal conversion.

e.g. 9 can be written in the power of 2 as,

$$9 = 2^3 + 2^2 + 2^1 + 2^0$$
$$= 8 + 0 + 0 + 1$$
$$\rightarrow 1001$$

Or, number 23 can be written as,

$$23 = 2^4 + 2^3 + 2^2 + 2^1 + 2^0$$
$$= 16 + 0 + 4 + 2 + 1$$
$$\rightarrow 10111$$

- A popular way to convert decimal to binary numbers is **double-dabble method** or **divide by two method.**

- In this method, progressively divide the decimal number by 2 and write down the remainders after each division.

- These remainders are taken in reverse order i.e. from bottom to top form the required binary number.

Example 1 : Convert the decimal number 25_{10} into equivalent binary number.

Solution :

	Quotient	Remainder	
$\dfrac{25}{2}$	12	1	LSB
$\dfrac{12}{2}$	6	0	
$\dfrac{6}{2}$	3	0	
$\dfrac{3}{2}$	1	1	Read
$\dfrac{1}{2}$	0	1	MSB

$$(25)_{10} = (11001)_2$$

It may also be put in the following form:

$$25 \div 2 = 12 + 1$$
$$12 \div 2 = 6 + 0$$
$$6 \div 2 = 3 + 0$$
$$3 \div 2 = 1 + 1$$
$$1 \div 2 = 0 + 1$$

$$1 \quad 1 \quad 0 \quad 0 \quad 1$$

$$(25)_{10} = (11001)_2$$

Example 2 : Convert the decimal number 45_{10} into equivalent binary number.

Solution :

	Quotient	Remainder
$\dfrac{45}{2}$	22	1 (LSB)
$\dfrac{22}{2}$	11	0
$\dfrac{11}{2}$	5	1
$\dfrac{5}{2}$	2	1 Read
$\dfrac{2}{2}$	1	0
$\dfrac{1}{2}$	0	1 (MSB)

$$(45)_{10} = (101101)_2$$

1.7 Decimal Fractions to Binary Conversion

- In this case, **multiply by two rule** is used. We multiply bit by two and record the carry in the integer position, these are taken in the forward direction (top to bottom) i.e. top carry is MSB and bottom is LSB.

Example 1 : Convert a decimal fraction   0.8125 into binary equivalent.

Solution :

$$0.8125 \times 2 = 1.625 \; = 0.625 \text{ with carry } 1 \quad \text{MSB}$$
$$0.625 \times 2 = 1.25 \; = \; 0.25 \text{ with carry } 1$$
$$0.25 \times 2 = 0.50 \; = \; 0.5 \text{ with carry } 0 \quad \text{Read}$$
$$0.5 \times 2 = 1.0 \; = \; 0 \text{ with carry } \quad 1 \quad \text{LSB}$$

Multiply the decimal fraction by 2 until the fractional product is zero.

$$(0.8125)_{10} = (0.1101)_2$$

Example 2 : Convert decimal number 25.6 into binary number.

Solution : First split the decimal number into an integer 25 and fraction of 0.6 and apply double-dabble to each part.

	Quotient	**Remainder**	
$\dfrac{25}{2}$	12	1	LSB
$\dfrac{12}{2}$	6	0	
$\dfrac{6}{2}$	3	0	Read
$\dfrac{3}{2}$	1	1	
$\dfrac{1}{2}$	0	1	MSB

$$(25)_{10} \ = \ (11001)_2$$

and

$0.6 \times 2 \ = \ 1.2 \ = \ 0.2$ with carry 1 MSB

$0.2 \times 2 \ = \ 0.4 \ = \ 0.4$ with carry 0

$0.4 \times 2 \ = \ 0.8 \ = \ 0.8$ with carry 0 Read

$0.8 \times 2 \ = \ 1.6 \ = \ 0.6$ with carry 1

$0.6 \times 2 \ = \ 1.2 \ = \ 0.2$ with carry 1 LSB

$(0.6)_{10} = (10011)_2$

In the fraction conversion, we stop the conversion process after getting five binary digits because in the last step, we had got the repetition of initial number i.e. 0.2.

$$(25.6)_{10} \ = \ (11001 \cdot 10011)_2$$

Example 3 : Perform the following : $(28.47)_{10} = (?)_2$.

Solution : For mixed numbers we have to perform conversion separately for integer part and fraction part.

2	28		
	14	0	
	7	0	
	3	1	$28_{10} = 11100_2$
	1	1	
	0	1	

$$0.47 \times 2 \;=\; 0.94 \;=\; 0.94 \qquad \text{with carry} \qquad 0$$

$$0.94 \times 2 \;=\; 1.88 = 0.88 \qquad \text{with carry} \qquad 1$$

$$0.88 \times 2 \;=\; 1.76 = 0.76 \qquad \text{with carry} \qquad 1$$

$$0.76 \times 2 \;=\; 1.52 = 0.52 \qquad \text{with carry} \qquad 1$$

$$\therefore \quad (0.47)_{10} \;=\; (0.0111)_2$$

$$\therefore \quad (28.47)_{10} \;=\; (11100.0111)_2$$

1.8 Hexadecimal Number System

- Hexadecimal number system has base of sixteen. It is extensively used in microprocessor work.

- They are much shorter than binary numbers. This makes them easy to write and remember.

- Hexadecimal number represents group of four bit binary numbers. It uses 16 distinct symbols 0 to 9 and A to F. Thus, the symbols in hexadecimal number systems are,

$$0, 1, 2, 3, 4, 5, 6, 7, 8, 9, \; A, B, C, D, E, F$$

- A subscript 16 or H is used which indicates a hexadecimal number.

    e.g.   42H or $(42)_{16}$;   9CH or $(9C)_{16}$

Table 1.4 : Hexadecimal number system

Decimal	Hexadecimal	Binary
0	0	0000
1	1	0001
2	2	0010
3	3	0011
4	4	0100
5	5	0101
6	6	0110
7	7	0111
8	8	1000
9	9	1001
10	A	1010
11	B	1011
12	C	1100
13	D	1101
14	E	1110
15	F	1111

- The hexadecimal number after F can be read as,

  10, 11, 12, 13, 14, 15, 16, 17, 18, 19, 1A, 1B, 1C, 1D, 1E, 1F,

  20, 21, ... 29, 2A, 2B, 2C, 2D, 2E, 2F,

  30, 31 .... so on ...

- With two hexadecimal digits we can count upto FF_{16} which is 255_{10}. To count beyond this, three hexadecimal digits are needed. For instance 100_{16} is decimal 256_{10}, 101_{16} is 257_{10} and so on. The maximum four digit hexadecimal number is $FFFF_{16}$ which is 65535_{10}.

1.9 Hexadecimal to Binary Conversion

- Hexadecimal number can be converted into binary number by converting each hexadecimal digit to 4 bit binary equivalent.

Example 1 : Convert $(9CA)_H$ to binary.

Solution :

$$
\begin{array}{ccc}
9 & C & A \\
\downarrow & \downarrow & \downarrow \\
1001 & 1100 & 1010
\end{array}
$$

$$(9CA)_H = (1001\ 1100\ 1010)_2$$

Example 2 : Convert $(D2E \cdot 8)_{16}$ to binary.

Solution :

$$
\begin{array}{ccccc}
D & 2 & E & \cdot & 8 \\
\downarrow & \downarrow & \downarrow & \downarrow & \downarrow \\
1101 & 0010 & 1110 & \cdot & 1000
\end{array}
$$

$$(D2E \cdot 8)_{16} = (1101\ 0010\ 1110 \cdot 1000)_2$$

1.10 Binary to Hexadecimal Conversion

- Binary number can be easily converted to hexadecimal by making a group of four bit starting from binary point. Then group is converted to equivalent hexadecimal.

Example 1 : Convert the 1100101001010111 binary number to hexadecimal.

Solution :

$$
\begin{array}{cccc}
1100 & 1010 & 0101 & 0111 \\
\downarrow & \downarrow & \downarrow & \downarrow \\
C & A & 5 & 7
\end{array}
$$

$$(1100101001010111)_2 = (CA57)_{16}$$

Example 2 : Convert $(10110110110011)_2$ to hexadecimal.

Solution :

$$\begin{array}{cccc} 0010 & 1101 & 1011 & 0011 \\ \downarrow & \downarrow & \downarrow & \downarrow \\ 2 & D & B & 3 \end{array}$$

$$(101101101100011)_2 \ = \ (2DB3)_{16}$$

Example 3 : Convert $11110101 \cdot 111$ to hexadecimal number.

Solution :

$$\begin{array}{cccc} 1111 & 0101 & \cdot & 1110 \\ \downarrow & \downarrow & & \downarrow \\ F & 5 & \cdot & E \end{array}$$

$$(1111\ 0101 \cdot 111)_2 \ = \ (F5 \cdot E)_{16}$$

1.11 Hexadecimal to Decimal Conversion

- In the hexadecimal number system each digit position corresponds to power of 16. The weights of the digit position in a hexadecimal number are as follows :

$$16^3\ 16^2\ 16^1\ 16^0 \cdot 16^{-1}\ 16^{-2}\ 16^{-3}$$
$$\uparrow$$
$$\text{Hexadecimal}$$
$$\text{point}$$

- Therefore, to convert from hexadecimal to decimal, multiply each hexadecimal digit by its weight and add the resulting products.

Example 1 : Convert $(E5)_{16}$ to decimal number.

Solution :
$$\begin{aligned} (E5)_{16} \ &= \ E \times 16^1 + 5 \times 16^0 \\ &= \ E \times 16 + 5 \times 1 \\ &= \ 14 \times 16 + 5 \times 1 \\ &= \ 224 + 5 \\ &= \ 229 \end{aligned}$$

$$(E5)_{16} \ = \ (229)_{10}$$

Example 2 : Convert $(B2F \cdot 39)_{16}$ to decimal number.

Solution :
$$\begin{aligned} (B2F \cdot 39)_{16} \ &= \ B \times 16^2 + 2 \times 16^1 + F \times 16^0 \\ &\quad + 3 \times 16^{-1} + 9 \times 16^{-2} \\ &= \ 11 \times 256 + 2 \times 16 + 15 \times 1 \\ &\quad + 0.1875 + 0.0352 \\ &= \ (2863 \cdot 2227)_{10} \end{aligned}$$

Example 3 : $(A2.2)_{16} = (?)_{10}$.

Solution : $\quad (A2.2)_{16} = A \times 16^1 + 2 \times 16^0 + 2 \times 16^{-1}$

$$= 10 \times 16 + 2 \times 1 + 2 \times \frac{1}{16}$$

$$= 160 + 2 + 0.125 = 162.125$$

$\therefore \quad (A2.2)_{16} = (162.125)_{10}$

1.12 Decimal to Hexadecimal Conversion

- The decimal number can be converted to hexadecimal number by successively dividing by 16. Then the remainder is converted to the equivalent hexadecimal. The remainder taken in reverse order, written from the bottom to the top, gives hexadecimal number.

Example 1 : Convert $(650)_{10}$ to hexadecimal by repeated division by 16.

Solution :

Successive division	Quotient	Remainder	Hexadecimal	
$\dfrac{650}{16}$	40	10	A	↑
$\dfrac{40}{16}$	2	8	8	Read
$\dfrac{2}{16}$	0	2	2	

$\therefore \qquad (650)_{10} = (28A)_{16}$

Example 2 : Convert $(423)_{10}$ to hexadecimal.

Solution :

Successive division	Quotient	Remainder	Hexadecimal	
$\dfrac{423}{16}$	26	7	7	↑
$\dfrac{26}{16}$	1	10	A	Read
$\dfrac{1}{16}$	0	1	1	

$\therefore \qquad (423)_{10} = (1A7)_{16}$

Example 3 : $(43.4)_{10} = (?)_{16}$.

Solution :

$$
\begin{array}{c|cc}
16 & 43 & \\
\hline
 & 2 & 11 \quad \rightarrow B \\
 & 0 & 2 \quad \rightarrow 2 \\
\end{array}
$$

$$(43)_{10} = (2B)_{16}$$

$0.4 \times 16 = 6.4 = 0.4$ with carry 6

$0.4 \times 16 = 6.4 = 0.4$ with carry 6

$0.4 \times 16 = 6.4 = 0.4$ with carry 6

$\therefore \qquad (43.4)_{10} = (2B.666)_{16}$

1.13 Hexadecimal Fraction

- For converting the decimal fraction to hexadecimal, multiply the decimal fraction by 16 and collect all the numbers to the left of the decimal point. Convert them into hexadecimal equivalent. Reading the number downwards gives the hexadecimal fraction.

Example 1 : Convert 0.357 to hexadecimal equivalent.

Solution : 0.357 is original number.

	Carry	Hexadecimal
$0.357 \times 16 = 5.712$	5	5
$0.712 \times 16 = 11.392$	11	B
$0.392 \times 16 = 6.272$	6	6

Reading the hexadecimal equivalent from top to the bottom.

$$(0.357)_{10} = (0.5B6)_{16}$$

1.14 Binary Coded Decimal (BCD)

- The BCD code expresses each digit in a decimal number by its 4 bit binary equivalent i.e. nibble. The 8421 code is a type of binary coded decimal (BCD) code and is composed of four bit having binary weight $(2^3 \, 2^2 \, 2^1 \, 2^0)$.

- With four bit combination 16 numbers can be represented. But in 8421 code only ten of these are used as given in Table 1.5.

Table 1.5

8421 (BCD)	Decimal
0000	0
0001	1
0010	2
0011	3
0100	4
0101	5
0110	6
0111	7
1000	8
1001	9

- The six code combinations 1010, 1011, 1100, 1101, 1110 and 1111 are invalid in the 8421 BCD code.

Example 1 : Convert each of the following decimal numbers into BCD.

Solution :

(i)   429

$$
\begin{array}{ccc}
4 & 2 & 9 \\
\downarrow & \downarrow & \downarrow \\
0100 & 0010 & 1001
\end{array}
$$

(ii)   9745

$$
\begin{array}{cccc}
9 & 7 & 4 & 5 \\
\downarrow & \downarrow & \downarrow & \downarrow \\
1001 & 0111 & 0100 & 0101
\end{array}
$$

(iii) 39.2

$$
\begin{array}{cccc}
3 & 9 & . & 2 \\
\downarrow & \downarrow & \downarrow & \downarrow \\
0011 & 1001 & . & 0010
\end{array}
$$

Example 2 : Find the decimal number represented by the following BCD code.

Solution : (i) 10000110

Make group of four bit

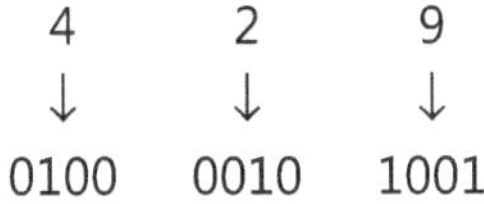

$$
\begin{array}{cc}
1000 & 0110 \\
\downarrow & \downarrow \\
8 & 6
\end{array}
$$

(ii)   100101110101

$$\begin{array}{ccc} 1001 & 0111 & 0101 \\ \downarrow & \downarrow & \downarrow \\ 9 & 7 & 5 \end{array}$$

1.14.1 Advantages and Disadvantages of BCD

Advantages :

* Decimal numbers can be easily converted into BCD numbers and vice versa. Only we have to remember the binary equivalent of decimal 0 through 9. Since we convert only one digit at a time.

Disadvantages :

* BCD numbers are not used in modern computers because of its limited value. A modern computer must be able to process alphanumeric. Therefore, computer uses binary numbers rather than BCD numbers.

* The rules for binary addition are not applicable to entire 8421 BCD number, but only to the individual 4 bit group. For example, consider the addition of 12 and 8.

Decimal	Binary	BCD	
12	1100	0001	0010
+ 08	+ 1000	+ 0000	1000
20	10100 ← 20	0001	1010

0001 1010 is not the BCD equivalent of 20. As 1010 does not exist in the BCD code.

1.15 Non-Weighted Codes

* Non-weighted binary codes do not follow the positional weighting principle, i.e., digit position within the number has not assigned fixed value. Examples of non-weighted code are gray code, excess-3 code and alphanumeric code etc.

* **Character Code:** In communication, along with ordinary number, we need some character, punctuation symbols and also control signals. The codes, we have studied so far, is not enough for data communication. For overcoming this problem, binary symbols are used for representing numbers, alphabetic character and special symbols.

* The binary codes which are used to represent number, alphabetic character and special symbols are called ***alphanumeric codes.***

- There are various codes which represent the alphanumeric data by using binary number. Some of them are :
  - (i) ASCII
  - (ii) EBCDIC.

1.16 ASCII
(American Code for Information Interchange)

- The ASCII code is extensively used for data communication and in computers. It is a 7-bit code, so it can represent 2^7 or 128 different characters.
- It can represent decimal number 0 to 9, letters of alphabet both upper and lower case, special symbols and control instructions.
- Each symbol has 7-bit code which is made up of a 3-bit group followed by a 4-bit group.
- The numbers are represented by 8421 BCD code preceded by 011. For instance, decimal number 3 is represented by 0110011.
- The Table 1.6 shows code combination of alphabate and other symbols. For example, upper case 'B' is represented by 1000010 whereas lower case 'b' is represented by 1100010, '+' is represented by 0101011.

Table 1.6 : American Standard Code for Information Interchange

		0	1	2	3	4	5	6	7
Row \ MSB →		000	001	010	011	100	101	110	111
Column	↓ LSB								
0	0000	NUL	DLE	SP	0	@	P		p
1	0001	SOH	DC_1	!	1	A	Q	a	q
2	0010	STX	DC_2	"	2	B	R	b	r
3	0011	ETX	DC_3	#	3	C	S	c	s
4	0100	EOT	DC_4	$	4	D	T	d	t
5	0101	ENQ	NAK	%	5	E	U	e	u
6	0110	ACK	SYN	&	6	F	V	f	v
7	0111	BEL	ETB	'	7	G	W	g	w
8	1000	BS	CAN	(	8	H	X	h	x
		0	1	2	3	4	5	6	7

Column	MSB →	000	001	010	011	100	101	110	111
Row	↓ LSB								
9	1001	HT	EM	)	9	I	Y	i	y
10	1010	LF	SUB	*	:	J	Z	j	z
11	1011	VT	ESC	+	;	K	[	k	{
12	1100	FF	FS	,	<	L	\	l	|
13	1101	CR	GS	–	=	M	]	m	}
14	1110	SO	RS	.	>	N	↑	n	~
15	1111	SI	US	/	?	O	←	o	DEL

Control Functions :

NUL	–	Null
SOH	–	Start of heading
STX	–	Start text
ETX	–	End text
EOT	–	End of transmission
ENQ	–	Enquiry
ACK	–	Acknowledge
BEL	–	Bell
BS	–	Backspace
HT	–	Horizontal tabulation
LF	–	Line feed
VT	–	Vertical tabulation
FF	–	Form feed
CR	–	Carriage return
SO	–	Shift out
SI	–	Shift in
DLE	–	Data link escape

DC_1, DC_2, DC_3, DC_4 –   Device control 1 – 4

NAK	–	Negative acknowledge

SYN	–	Synchronous idle
ETB	–	End of transmission block
CAN	–	Cancel
EM	–	End of medium
SUB	–	Substitute
ESC	–	Escape
FS	–	File separator
GS	–	Group separator
RS	–	Record separator
US	–	Unit separator
DEL	–	Delete

1.17 Gray Code

- The gray code is an unweighted code, which means that there is no specific weight assigned to the bit position. Each gray-code number differs from the preceding number by a single bit.

Table 1.7: Gray code

Decimal	Gray code	Binary
0	0000	0000
1	0001	0001
2	0011	0010
3	0010	0011
4	0110	0100
5	0111	0101
6	0101	0110
7	0100	0111
8	1100	1000
9	1101	1001
10	1111	1010
11	1110	1011
12	1010	1100
13	1011	1101
14	1001	1110
15	1000	1111

- The Table 1.7 shows listing of gray code number for decimal and binary from 0 to 15. Gray code can have any number of bit.

- Considering gray code from decimal 3 to 4, the gray code changes from 0010 to 0110 while the binary code for the same i.e. 3 to 4 changes from 0011 to 0100.

Binary to Gray Conversion:

- The MSB of binary is copied as it is, this represents the MSB of gray code. Then going from left to right add each adjacent pair of binary digits to get the next gray code digit. In addition, carry (if any) is ignored.

Example 1 : Find gray code of binary number 1101.

Solution :

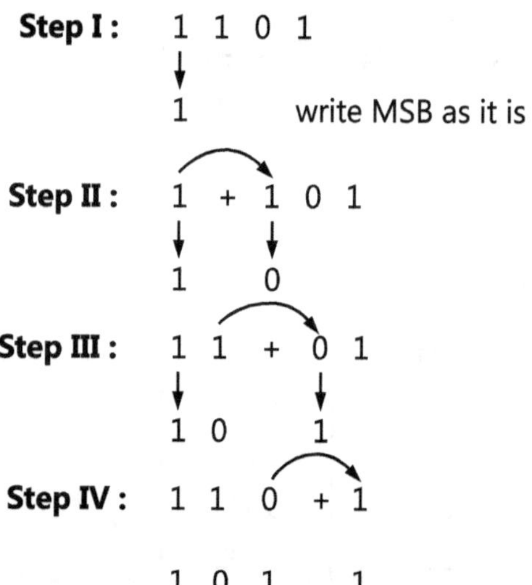

Or we may write the same example as 1101

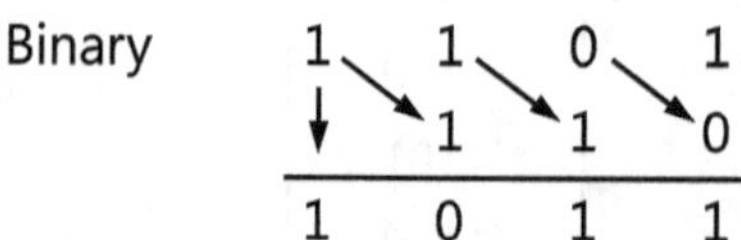

The gray code for binary 1101 is 1011.

Example 2 : Convert the following binary numbers to the gray code :

(i) 1100_2,   (ii) 1010_2,   (iii) 11010_2,   (iv) 101110_2.

Solution :

(i)

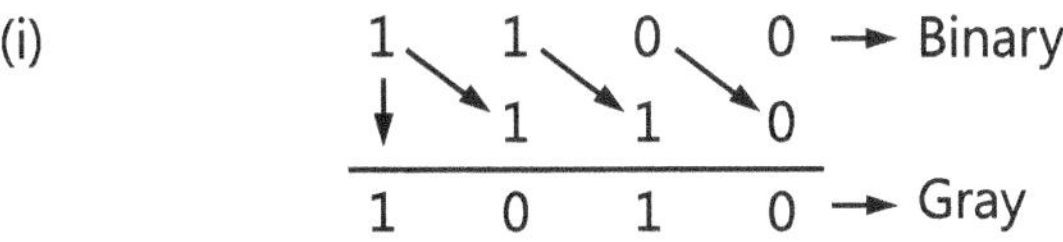

(ii) 1010_2

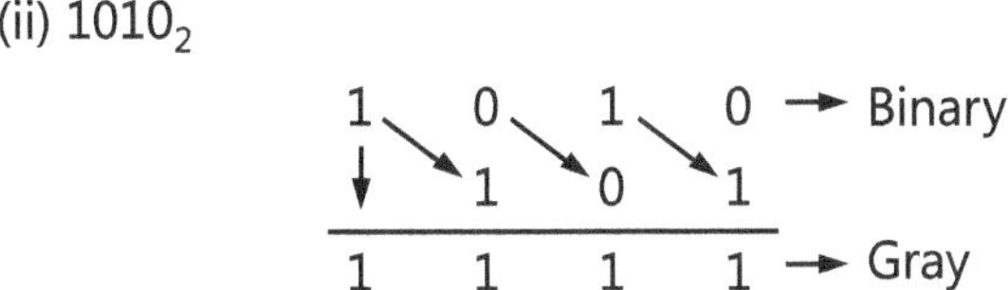

(iii) 11010_2

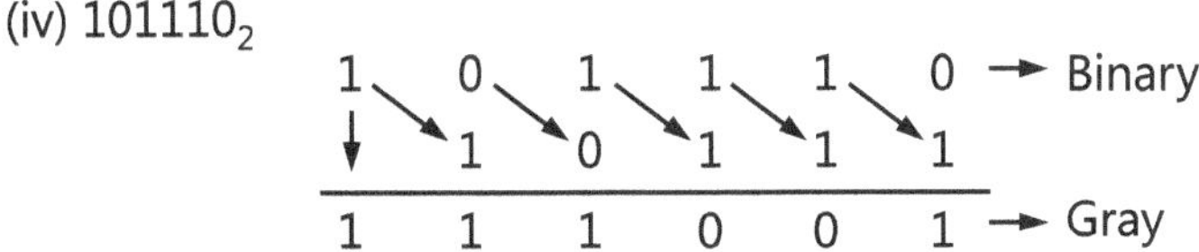

(iv) 101110_2

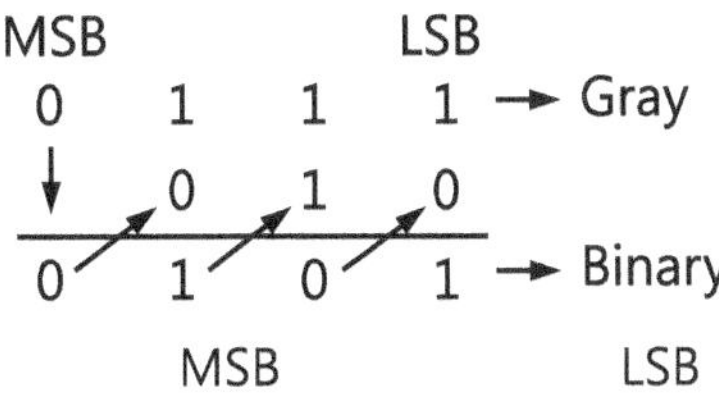

Gray Code to Binary Conversion :

- To convert gray code to binary, first write the MSB (left most bit) same as corresponding digit in gray code and then add each MSB of binary to the next bit of gray code. If carry is generated, it should be ignored. Process is carried out till LSB is reached.

Example 1 : Convert gray code number 0111 to binary.

Solution :

i.e. Gray code 0111 has binary 0101.

Solved Examples

Example 1.1 : *Solve the following :*

(i) $(99)_{10} = (?)_{16}$

Solution : For conversion from decimal to hexadecimal, decimal number is divided by 16 (base of hexadecimal).

Successive division	Quotient	Remainder	
$\dfrac{99}{16}$	6	3	
$\dfrac{6}{16}$	0	6	↑ read

Reading the remainder from bottom to the top.

$$(99)_{10} = (63)_{16} \quad \text{or} \quad 63H$$

(ii) $(1011101)_2 = (?)_{16}$

Solution : Binary numbers can be converted into the equivalent hexadecimal number by making groups of four bit starting from LSB and moving towards MSB. Then replace each group of four bit by its hexadecimal representation.

Start from LSB.   1011101

$$\underbrace{0\ 1\ 0\ 1} \quad \underbrace{1\ 1\ 0\ 1}$$
$$\downarrow \qquad \downarrow$$
$$5 \qquad\quad D$$

$\therefore$ $(1011101)_2 = (5D)_{16} \quad \text{or} \quad 5DH$

H stands for hexadecimal.

(iii) $(FA)_{16} = (?)_{10}$

Solution : To convert hexadecimal to decimal, multiply each hexadecimal digit by its weight and add the resulting product.

$$FA = F \times 16^1 + A \times 16^0$$
$$= 15 \times 16^1 + 10 \times 16^0$$
$$= 240 + 10 = 250$$

$\therefore$ $(FA)_{16} = (250)_{10}$

OR Another method can be used to evaluate a hexadecimal number in terms of its decimal equivalent, first convert the hexadecimal number to binary and then convert from binary to decimal by following way.

$\therefore$ $(FA)_{16}$

First convert FA into binary form

$$F \qquad A$$
$$\downarrow \qquad \downarrow$$
$$1111 \qquad 1010_2$$

$\therefore \qquad 11111010_2 = 2^7 \times 1 + 2^6 \times 1 + 2^5 \times 1 + 2^4 \times 1$

$$+ \ 2^3 \times 1 + 2^2 \times 0 + 2^1 \times 1 + 2^0 \times 0$$
$$= \ 128 + 64 + 32 + 16 + 8 + 0 + 2 + 0$$
$$= \ 128 + 64 + 32 + 16 + 8 + 2 = 250_{10}$$

$\therefore \qquad (FA)_{16} = (250)_{10}$

Example 1.2 : *Convert* $(76)_{10}$ *into* *(i) Binary, (ii) Hexadecimal.*

Solution : (i) Binary : Convert into binary form by repeated division of 2.

Successive division	Quotient	Remainder	
$\dfrac{76}{2}$	38	0	
$\dfrac{38}{2}$	19	0	
$\dfrac{19}{2}$	9	1	
$\dfrac{9}{2}$	4	1	
$\dfrac{4}{2}$	2	0	
$\dfrac{2}{2}$	1	0	$\uparrow$
$\dfrac{1}{2}$	0	1	read

Reading from bottom to the top,

$$(76)_{10} = (1001100)_2$$

(ii) Hexadecimal :

Successive division	Quotient	Remainder	
$\dfrac{76}{16}$	4	$12 \rightarrow C$	
$\dfrac{4}{16}$	0	4	$\uparrow$
			read

Reading the remainders from bottom to the top,

$$(76)_{10} \ = \ (4C)_{16}$$

Example 1.3 : *Convert* $(26.457)_{10} \ = \ (?)_2$

Solution : Double-dabble method can be used to convert decimal to binary number.

Let us convert decimal number 26 to its binary equivalent.

$\dfrac{26}{2} = 13$ with remainder of 0

$\dfrac{13}{2} = 6$ with remainder of 1

$\dfrac{6}{2} = 3$ with remainder of 0

$\dfrac{3}{2} = 1$ with remainder of 1

$\dfrac{1}{2} = 0$ with remainder of 1   Read

The remainders taken in reverse order (from bottom to the top)

$$(26)_{10} \ = \ (11010)_2$$

As far as fractions are concerned, multiply by number 2 and record a carry in the integer position. The carries taken in forward order are the binary fraction.

Convert   0.457 to binary fraction,

$0.457 \times 2 = 0.914 = 0.914$ with carry   0

$0.914 \times 2 = 1.828 = 0.828$ with carry   1

$0.828 \times 2 = 1.656 = 0.656$ with carry   1

$0.656 \times 2 = 1.312 = 0.312$ with carry   1

$0.312 \times 2 = 0.624 = 0.624$ with carry   0

$0.624 \times 2 = 1.248 = 0.248$ with carry   1

Read

Taking the carries in forward order give binary fraction 0.011101. In this case, we stop the process after getting 6 binary digits. This answer gives an approximation. If more accuracy is needed, continue multiplying by 2 until you have as many digits as necessary.

Example 1.4 : *Convert $(8.5625)_{10}$ to binary.*

Solution : For mixed decimal (number with integer and fractional part) we have to convert integer part and fractional part separately into binary.

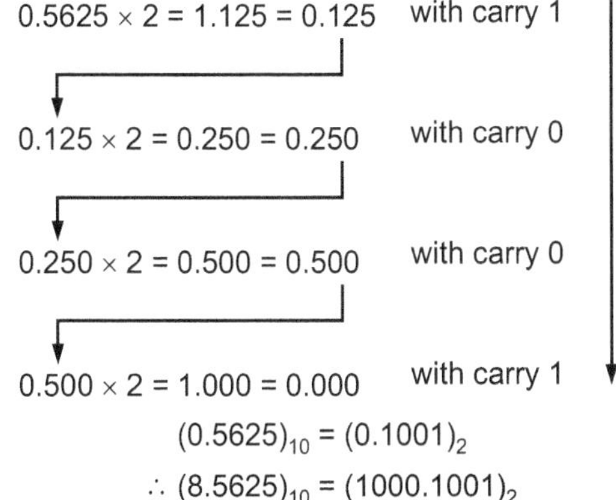

$8_{10} = 1000_2$

Convert decimal fraction 0.5625 into binary

$0.5625 \times 2 = 1.125 = 0.125$ with carry 1

$0.125 \times 2 = 0.250 = 0.250$ with carry 0

$0.250 \times 2 = 0.500 = 0.500$ with carry 0

$0.500 \times 2 = 1.000 = 0.000$ with carry 1

$$(0.5625)_{10} = (0.1001)_2$$

$$\therefore (8.5625)_{10} = (1000.1001)_2$$

1.18 Rules of Binary Addition and Subtraction

1.18.1 Binary Addition

- There are four basic rules for adding binary digits :

$$0 + 0 = 0$$
$$0 + 1 = 1$$
$$1 + 0 = 1$$
$$1 + 1 = 10_2$$

Table 1.8 : Rules of binary addition

Augend	Addend	Sum	Carry	Result
0	0	0	0	0
0	1	1	0	1
1	0	1	0	1
1	1	0	1	10

- In the first three rows in Table 3.1, there is no carry i.e. carry $= 0$, whereas in the fourth row a carry is produced i.e. carry $= 1$ and it is added to the next higher binary position as similar to decimal addition.

- When there is a carry, we have a situation where three bits are being added (a bit in each of two numbers and a carry bit). The rules for this are as follows :

$$\text{Carry bits} \begin{cases} 1 + 0 + 0 = 01_2 & 1 \text{ with a carry of } 0 \\ 1 + 1 + 0 = 10_2 & 0 \text{ with a carry of } 1 \\ 1 + 0 + 1 = 10_2 & 0 \text{ with a carry of } 1 \\ 1 + 1 + 1 = 11_2 & 1 \text{ with a carry of } 1 \end{cases}$$

- The subscript 2 (like 10_2) represents a binary number (8 for octal, 10 for decimal, 16 for hexadecimal).

Example 1 : (a)

$$
\begin{array}{r}
01_2 \\
+\ 01_2 \\
\hline
10_2
\end{array}
\qquad\qquad
\begin{array}{r}
01_{10} \\
+\ 01_{10} \\
\hline
02_{10}
\end{array}
$$

(b)

$$
\begin{array}{r}
11_2 \\
+\ 11_2 \\
\hline
110_2
\end{array}
\qquad\qquad
\begin{array}{r}
3_{10} \\
+\ 3_{10} \\
\hline
6_{10}
\end{array}
$$

(c)
$$111_2 \qquad\qquad 7_{10}$$
$$\underline{+\ 11_2} \qquad\qquad \underline{+\ 3_{10}}$$
$$1010_2 \qquad\qquad 10_{10}$$

(d)
$$1011$$
$$\underline{+\ 1100}$$
$$10111$$
$$\uparrow$$
carry

(e)   Add two numbers 0101 and 1111.

$$
\begin{array}{ccccc}
(1) & (1) & (1) & & \leftarrow \text{carry} \\
0 & 1 & 0 & 1 & \\
+\ 1 & 1 & 1 & 1 & \\
\hline
1\ \ 0 & 1 & 0 & 0 & \\
\end{array}
$$
$$\uparrow$$
carry

For 8-bit arithmetic addition, least significant bit (LSB) is on the right and most significant bit (MSB) is on the left.

(MSB)   (LSB)

$$A_7\ A_6\ A_5\ A_4\ A_3\ A_2\ A_1\ A_0$$

$$B_7\ B_6\ B_5\ B_4\ B_3\ B_2\ B_1\ B_0$$

Example 2 : Add two 8-bit numbers 01010111 and 00110101.

Sol. :   (1) (1) (1) (1) (1) (1) $\leftarrow$ carry

$$
\begin{array}{l}
\ \ 0\ 1\ 0\ 1\ 0\ 1\ 1\ 1 \\
+\ 0\ 0\ 1\ 1\ 0\ 1\ 0\ 1 \\
\hline
\ \ 1\ 0\ 0\ 0\ 1\ 1\ 0\ 0 \\
\end{array}
$$

1.18.2 Binary Subtraction

- There are four basic rules for subtracting binary digits.

$$0 - 0\ =\ 0$$
$$1 - 1\ =\ 0$$
$$1 - 0\ =\ 1$$
$$10_2 - 1\ =\ 1$$

Table 1.9 : Rules of binary subtraction

Minuend	Subtrahend	Difference	Borrow
0	0	0	0
0	1	1	1
1	0	1	0
1	1	0	0

- While subtracting numbers, we sometime have to borrow from the next higher column. In Table 1.9 except in the second row, borrow = 0. When borrow = 1, as in the second row, this is to be subtracted from the next higher binary bit as it is done in decimal subtraction.

Example 1 : (a)

$$11_2 \qquad\qquad 3_{10}$$
$$\underline{-\ 10_2} \qquad\qquad \underline{-\ 2_{10}}$$
$$01_2 \qquad\qquad 1_{10}$$

(b) Subtract 011 from 101.

When 1 is borrowed, 0 is left

```
      0                    Borrow 1 from next column making
    1  10  1               10 in this column (10 – 1 = 1)

  -  0  1  1

     0  1  0               ← (1 – 1 = 0)
```

(c) Subtract the following :

$$1011$$
$$\underline{-\,0110}$$
$$0101$$

In columns 1 and 2, borrow = 0 and in column 3, it is 1. Therefore, in column 4, first subtract 0 from 1 and from this result obtained, subtract the borrow bit.

1.19 Subtraction Using 1's Complement Method

- The 1's complement of binary number is found by simply changing all 1s to 0s and all 0s to 1s as illustrated by examples.

Binary number	**1's complement**
11010	00101
01001	10110
111100	000011
01011010	10100101

1's Complement Subtraction

- To subtract a smaller number from a larger number, the 1's complement method is as follows :

1.   Determine the 1's complement of the smaller number.

2.   Add the 1's complement to the larger number.

3.   Remove the carry and add it to the result. This is called end-around carry.

Example 1 : Subtract 1001 from 1100 using the 1's complement method.

Solution :

Direct subtraction   **1's complement method**

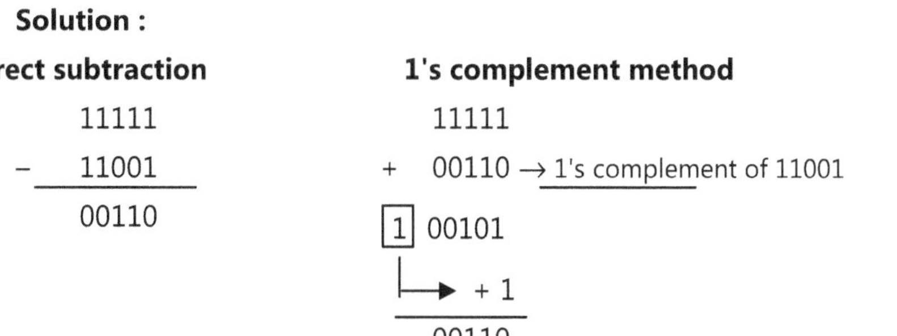

Example 2 : Subtract 11001 from 11111 using 1's complement method.

Solution :

Direct subtraction   **1's complement method**

```
        11111                           11111
   -    11001                      +     00110 → 1's complement of 11001
        00110                         1  00101
                                           + 1
                                         00110
```

Example 3 : Subtract 1000 from 0111 using 1's complement method.

Solution :

	Direct subtraction	**1's complement method**

<table>
<tr><td>7</td><td>0111</td><td>0111</td></tr>
<tr><td>– 8</td><td>– 1000</td><td>0111 → 1's complement</td></tr>
<tr><td>– 1</td><td>– 0001</td><td>1110 ← There is no carry,</td></tr>
</table>

For direct subtraction, subtract smaller number from larger number and assign the sign of larger number.

∴ result is –ve and is in 1's complement form.

∴ result is – 0001.

1.20 Subtraction Using 2's Complement Method

- The 2's complement of a binary number is found by adding 1 to 1's complement.

Example 1 : Binary number is 1010. 1's complement of 1010 is 0101. Add 1 to 0101

$$i.e. \quad 0101$$
$$\underline{+\ 1}$$
$$0110$$

Example 2 : Binary number 10101 has 2's complement as,

$$01010 \quad \text{1's complement of 10101}$$
$$\underline{+\ 1 \quad \text{Add 1}}$$
$$01011 \quad \text{2's complement of 10101}$$

2's Complement Subtraction

- To subtract a smaller number from larger one, 2's complement method is applied as follows :

1. Determine the 2's complement of the smaller number.
2. Add the 2's complement to the larger number.
3. Discard the carry (there is always a carry in this case).

Example 1 : Subtract 1101 from 1111 using 2's complement method.

Solution : Direct subtraction :

$$1111$$
$$\underline{-\ 1101}$$
$$0010$$

2's complement method :

$$1111$$

$$+ \; 0011 \quad \rightarrow \; \text{2's complement of 1101}$$

$$\boxed{1} \;\; 0010$$

$$\uparrow$$

discard the carry

Final result is 0010.

Example 2 : Perform the following subtraction.

Solution : Direct method :

$$1110 \qquad\qquad 1110$$

$$- \; 1001 \qquad\qquad + \; 0111 \quad \rightarrow \text{2's complement of 1001}$$

$$0101 \qquad\qquad \boxed{1} \;\; 0101$$

$$\uparrow$$

discard the carry $\therefore$ **Result :** 0101

Example 3 : Perform subtraction 11000 from 11111.

Solution : Direct method :

$$11111 \qquad\qquad 11111$$

$$- \; 11000 \qquad\qquad + \quad 01000 \quad \rightarrow \text{2's complement of 11000}$$

$$00111 \qquad\qquad \boxed{1} \;\; 00111$$

$$\uparrow$$

discard the carry $\therefore$ **Result :** 00111

$$11111_2 \qquad\qquad\qquad 31_{10}$$

$$- \; 11000_2 \qquad\qquad\; - \; 24_{10}$$

$$00111_2 \qquad\qquad\quad - \; 07_{10}$$

$$\downarrow \qquad\qquad\qquad\quad \downarrow$$

Binary subtraction   Decimal subtraction

1.21 Introduction to Logic Gates

- Digital signal has two discrete levels or values 1 and 0, and are referred as HIGH (ON) and LOW (OFF) states.

- A transistor operates very reliably in the switching mode. So transistor is one of the main building blocks of a digital circuit.

- The simplest way to use a transistor is a switch, meaning that we operate it at either saturation or cut-off. When saturated, a transistor is like a closed switch, whereas cut-off is like an open switch.
- A diode can also be used as a switch. In forward bias condition, resistance of diode is zero and it behaves like closed switch. On the other hand, under reverse bias, the resistance is very high and no current can flow, which makes it as an open switch.
- The logic gate has one or more inputs. Depending upon the inputs it produces a HIGH (1) or LOW (0). A gate receives binary input and produces output which depends upon the inputs and the function intended for.

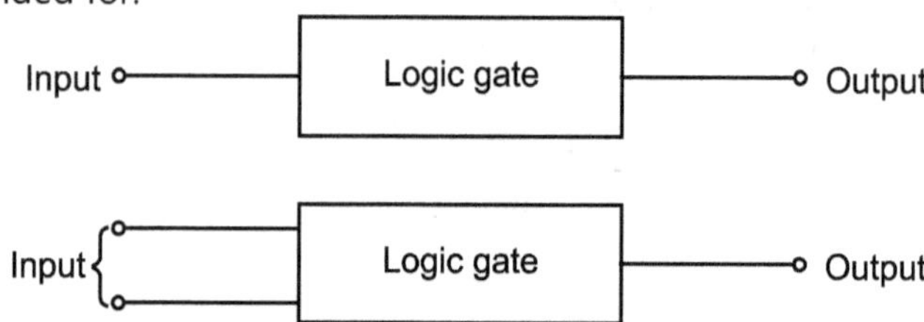

Fig. 1.3: Block diagram of logic gate

- For simplicity in designing and understanding of digital system, each gate has assigned symbol. There are many types of symbols depending on function they perform. The most commonly used symbols are given in Table 1.10.

Table 1.10 : Logic gates

Sr. No.	Logic gate	Function	Symbol
1.	Inverter	NOT	A — Y
2.	AND	AND	A, B — Y
3.	NAND	NAND	A, B — Y
4.	OR	OR	A, B — Y
5.	NOR	NOR	A, B — Y
6.	Exclusive OR	EX-OR	A, B — Y
7.	Exclusive NOR	EX-NOR	A, B — Y

1.22 Positive and Negative Logic

- In a digital system there are two discrete levels HIGH (1) and LOW (0). If the higher of the two voltages represents a 1 and lower voltage represents 0, the system is called ***positive logic*** system.
- On the other hand, if lower voltage represents a 1 and higher voltage represents 0 we have ***negative logic*** system.
- Suppose that +5V and 0V are our logic level voltages. We will designate +5V as HIGH (1) and 0V as LOW (0). So positive and negative logic can be defined as :

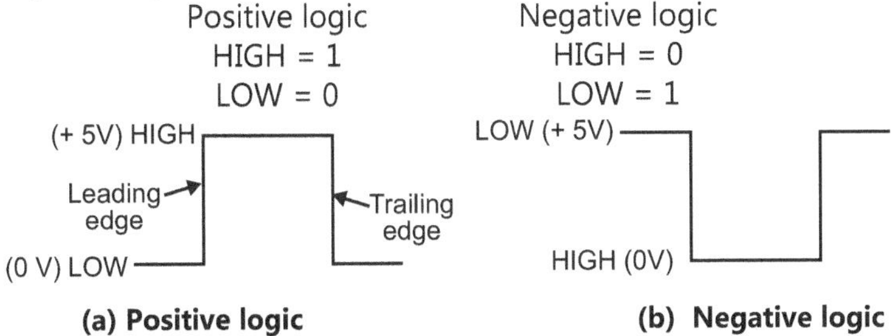

(a) Positive logic   **(b) Negative logic**

Fig. 1.4 : Digital signal representation

- From Fig. 1.4 (a), + 5V will be considered as HIGH level in positive logic and LOW level in negative logic system.
- Similarly, voltage level 0V will be considered as LOW in the positive logic system and HIGH level in the negative logic system, as in Fig. 1.4 (b).

1.23 Basic Logic Gates
(Universal and Derived Gates)

- In a digital system, there are only few basic operations performed irrespective of the complexities of the system. These operations may be required to be performed number of times in a digital computer or digital control system.
- The basic logic gates are AND, OR, NOT etc. These basic gates can be combined to perform other important logic operations like NAND, NOR and EX-NOR gates. So these are called as ***derived gates.***
- Any Boolean (or logic) expression can be realized by using the AND, OR and NOT gates. NAND, NOR operations can be derived from it. These operations have become very popular and are widely used because either NAND, NOR gates are sufficient for the realization of any logical expression. Because of this reason, NAND and NOR gates are known as universal gates.

1.24 AND Gate

- The AND gate performs logical multiplication, more commonly known as AND function.

- The AND gate is composed of two or more inputs and single output. The output of AND gate is high (1) only when all the inputs are high (1).

- A logic gate can be used as diode, transistor, FET or combination of these elements.

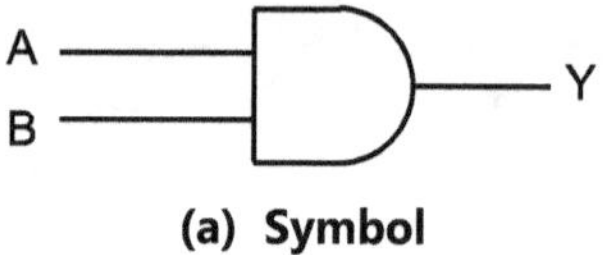

(a) Symbol

- The inputs of the two input AND gates are labelled as A and B and output is labelled as Y.

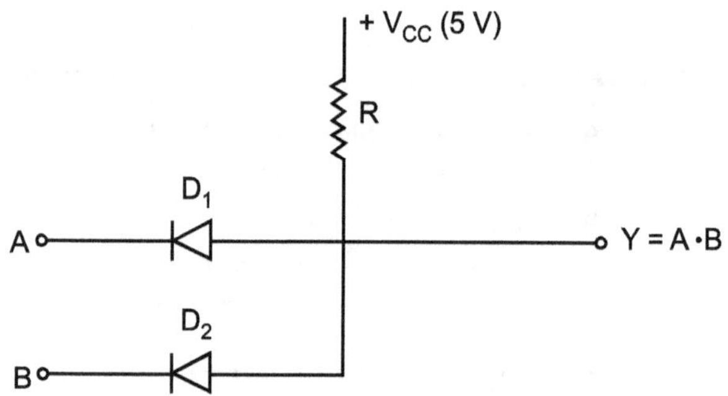

(b) Simplified circuit with diode

Inputs		Output
A	**B**	**Y**
0	0	0
0	1	0
1	0	0
1	1	1

(c) Truth table

Fig. 1.5

- The AND gate can be designed using diodes. Fig. 1.5 (b) shows two input AND gates using diodes. There are only four possible input cases.

Case I : A is low and B is low. With this situation both diodes D_1 and D_2 are forward bias by supply voltage and will conduct. Because of this, the output voltage is ideally zero. This means Y is low. i.e. Y = 0

Case II : A is low and B is high since diode D_1 is forward bias pulling the output down to a low voltage. The diode D_2 is reversed bias. The output Y = 0.

Case III : A is high and B is low. The diode D_2 is forward bias and pulling the output down to a low voltage. The diode D_1 is reversed bias, therefore Y = 0.

Case IV : A is high and B is high with both inputs at + 5V. Both diodes are non-conducting because the voltage across each is zero. Therefore, current will not flow through R and output Y is at high (logic 1) i.e. Y = 1.

1.25 OR Gate

- An OR gate performs logical addition, more commonly known as OR function. It has two or more inputs and one output. The logic symbol of OR gate is shown in Fig. 1.6 (a).

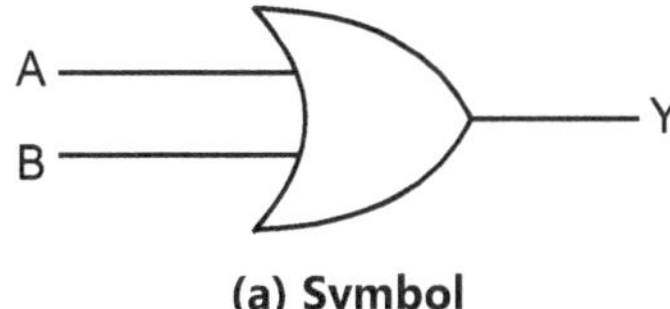

(a) Symbol

- The inputs of two input OR gates are labelled as A and B. Output is labelled as Y.

- The operation of OR gate is such that a HIGH on the output is produced when any of the inputs are HIGH. Output is LOW only when all of the inputs are LOW. In certain situation, if we want a output HIGH, when one or more of its inputs is HIGH, OR gate can be used.

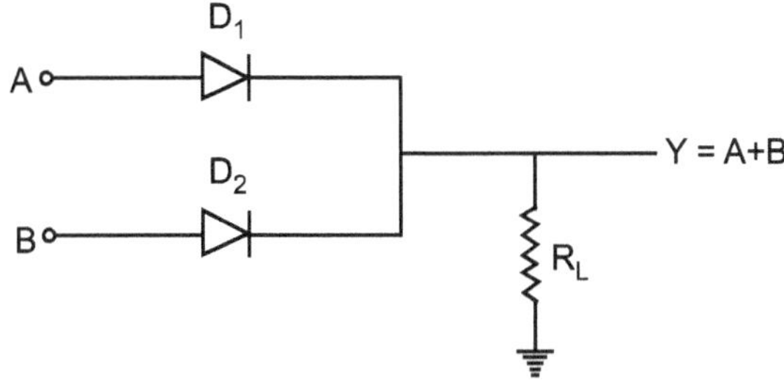

(b) Diode circuit of OR gate

Inputs		Output
A	**B**	**Y = A + B**
0	0	0
0	1	1
1	0	1
1	1	1

(c) Truth table

Fig. 1.6

Let us assume input voltage 1 (High state) or + 5V and 0 (Low state) is 0 V.

For different input combinations, we will have the following cases :

Case I : A and B both LOW i.e. A = 0 and B = 0. In this case, both diodes D_1 and D_2 are non-conducting, therefore, Y output is LOW. i.e. Y = 0.

Case II : A is LOW and B is HIGH i.e. A = 0 and B = 1.

The high B input voltage of the diode D_2 will conduct as it is under forward bias condition. The conducting current will flow through resistor R_L and high voltage will develop at the output. This will make Y = 1 (High logic). The diode D_1 is reversed bias.

Case III : A is HIGH and B is LOW i.e. A = 1 and B = 0.

The diode D_1 will conduct and D_2 is non-conducting. The conducting current in this case also flows through resistor R_L and HIGH voltage will develop at the output. This will make Y = 1 (High logic).

Case IV : A and B both are HIGH i.e. A = 1 and B = 1. Diodes D_1 and D_2 will conduct and therefore output is at high logic level i.e. Y = 1.

1.26 NOT Gate

- The NOT gate is also called *inverter* as it inverts the input signal. The inverter (NOT circuit) performs a basic logic function called inversion or complementation. It is called NOT because its output is NOT the same as input. The purpose of NOT or inverter is to change one logic level to the opposite level. In terms of bits, it changes 1 to 0 and 0 to 1.

(a)   **(b)**

Input A	Output Y = $\overline{A}$
0	1
1	0

(c) Truth table

Fig. 1.7 : NOT gate

- Fig. 1.7 (a) and (b) shows standard symbols for NOT gate. As shown in Fig. 1.7, NOT gate has one input A and one output Y. Its logic expression is,

$$Y = NOT\ A = \overline{A}$$

and is read as "Y equals NOT A" or Y equals complement of A.

If   A = 0, $Y = NOT\ 0 = 1$

On the other hand,

If   A = 1, $Y = NOT\ 1 = 0$

- The presence of small circle, known as the bubble, always denotes inversion in digital circuits.

1.27 Transistor Inverter

- Transistor has a property to invert an input signal, so a transistor can be used as an inverter.

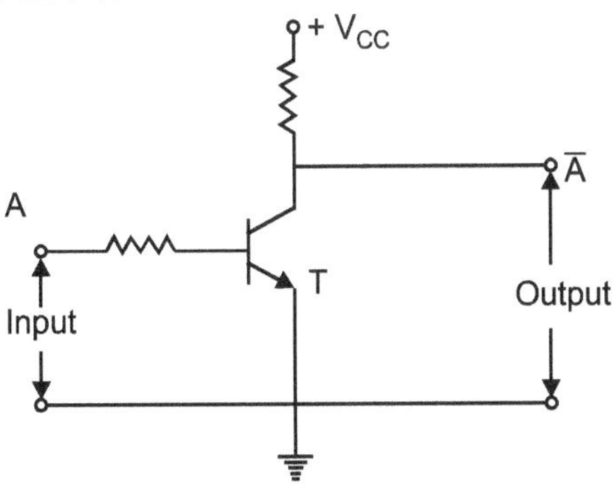

Fig. 1.8 : Transistorized NOT gate

- The Fig. 1.8 shows transistor as an inverter. When the voltage applied at A is + 5V or 1, the transistor will be fully turned ON and the output which is taken at the collector, will become very low (logic (0)).

- On the other hand, when the applied input voltage at A is 0V or logic 0, the transistor will be cut-off making the output voltage high. Obviously, in each case, the output is inverse of the input.

- If two inverters are cascaded as shown in Fig. 1.9, the circuit will behave like a non-inverting amplifier i.e. output will be same as that of input.

Fig. 1.9 : Buffer

- The buffer is basically used to isolate two other devices. A buffer has a high input and a low output impedance. This results in a low input current and high output current.

1.28 Universal and Derived Gates

- Any Boolean (or logic) expression can be realized by using the AND, OR and NOT gate as discussed before. From these three operations, two more operations have to be derived: the NAND and NOR operations. These operations are very popular and are widely used.

- NAND and NOR gates are known as *universal gates* because these can perform any logical function of AND, OR and NOT gate. NAND or NOR gates are sufficient for the realization of any logical expression.

(a) NAND gate :

- The NOT-AND operation is known as the NAND operation. It can be obtained by connecting a NOT gate in the output of an AND gate as it gives the output in the form of logic expression: $Y = \overline{A \cdot B}$

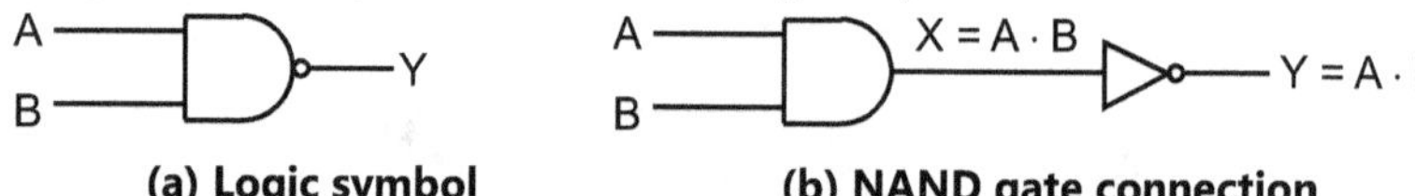

(a) Logic symbol   (b) NAND gate connection

- This gate gives an output 1 if its both inputs are not 1. In other words, it gives an output 1 if either A or B or both are 0.

Inputs		Outputs	
A	**B**	**AND X = A·B**	**NAND Y = $\overline{A \cdot B}$**
0	0	0	1
0	1	0	1
1	0	0	1
1	1	1	0

(c) Truth table

Fig. 1.10 : NAND gate

- The truth table for two input NAND gates is given in Fig. 1.10 (c). It is just opposite of the truth table for AND gate.
- The diode-transistor equivalent of a NAND gate is shown in Fig. 1.11.

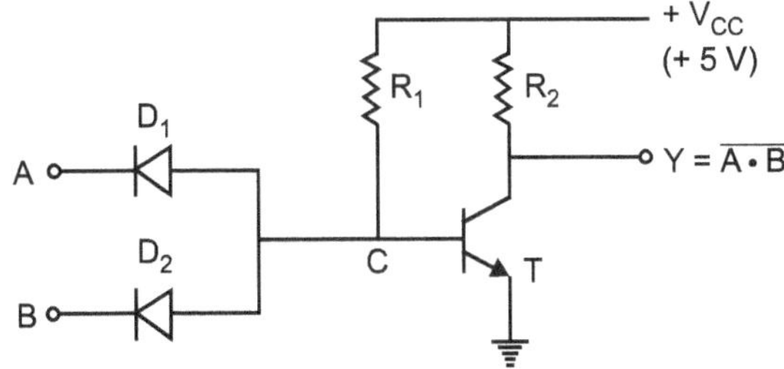

Fig. 1.11: NAND equivalent circuit

- It is seen that, point C would be driven to ground when either D_1 or D_2 or both D_1 and D_2 conduct. It represents input conditions from truth table, AB = 00, 01 and 10. Transistor is cut-off and hence Y output goes to + V_{CC} or logic level 1.
- When AB = 11 i.e. A = 1 and B = 1 the point C is at + 5V and transistor goes saturation, output becomes 0 V or logic level 0.

(b) NOR gate:

- It is a combination of NOT OR gate. It can be made by connecting an inverter to the output of OR gate.
- Output of NOR gate will be complement of output of OR gate. The logical operation of NOR gate is such that a LOW output occurs when any of its inputs are HIGH. It has HIGH output only when all inputs are LOW.

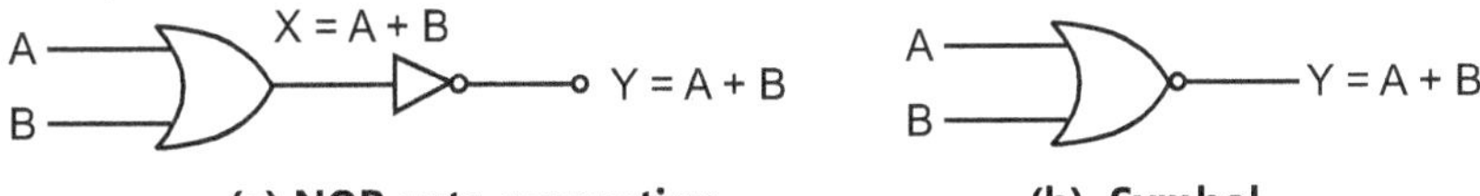

(a) NOR gate connection   **(b) Symbol**

Inputs		Outputs		
A	**B**	**OR** X = A + B		**NOR** Y $= \overline{A + B}$
0	0	0		1
0	1	1		0
1	0	1		0
1	1	1		0

(c) Truth table

Fig. 1.12 : NOR gate

From Fig. 1.12 (c) inputs are labelled as A and B and output is labelled as Y.

$$\therefore \qquad Y = \overline{A + B}$$

Note that, this operation results in opposite to that of OR gate.

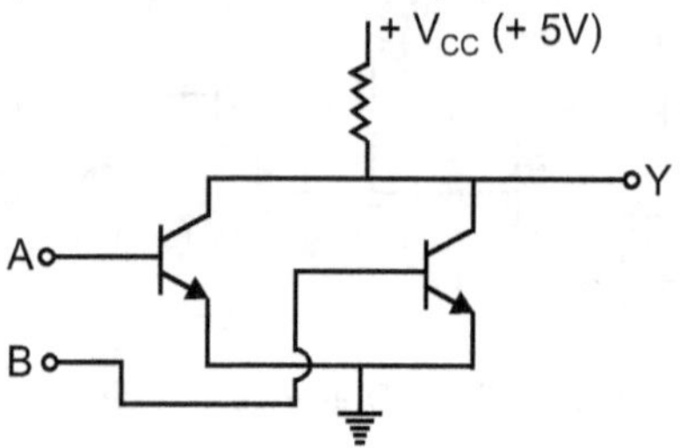

Fig. 1.13 : Equivalent circuit for NOR gate

- A transistor equivalent circuit of NOR gate is shown in Fig. 1.13. As seen, Y is 1 only when both transistors are cut-off i.e. A = 0 and B = 0. For any other input condition like, AB = 01, 10 and 11 both transistors saturate forcing point Y to goto LOW state.

1.29 Transistor or Gate

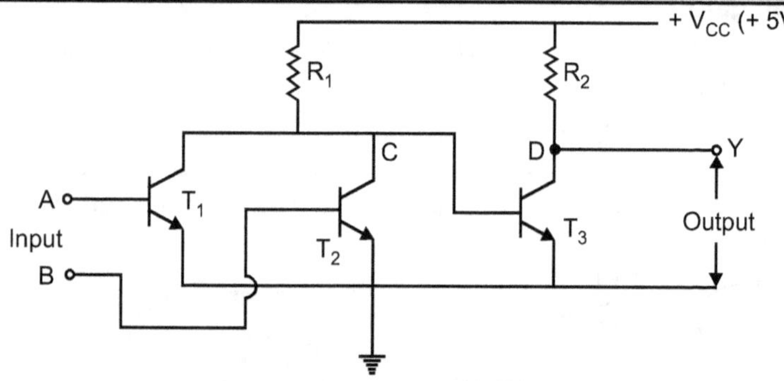

Fig. 1.14 : Transistor OR gate

- As we know OR gate has output 1 (high logic) even when one of the inputs is high (logic 1).
- Possible transistor OR gate consists of three transistors T_1, T_2 and T_3 supplied from a common supply $+V_{CC} = + 5V$.

Case I : When + 5V is applied to A, transistor T_1 is forward biased and conducts. Assuming that T_1 is saturated, entire $+ V_{CC} = + 5V$ drops across R_1 thus causing point C to go to ground. This in turn, cut-off transistor T_3 and causes output Y to go to V_{CC} i.e. + 5V or logic 1. i.e. A = 1, output Y = 1.

Case II : When + 5V is applied to B, transistor T_2 conducts or forward bias, thereby driving point C to ground i.e. 0 V. The base of transistor T_3 is at 0 V, so T_3 is cut-off. Thus, driving output Y again to + V_{CC} (i.e. + 5V) or logic 1 (i.e. B = 1), output Y = 1.

Case III : When A = 1 and B = 1, it works as discussed in cases I and II i.e. output Y = 1 (High).

Case IV : When both inputs A and B are grounded i.e. 0 V, transistors T_1 and T_2 are cut-off driving point C to + 5V. As a result, transistor T_3 becomes forward biased and conducts. In this case, entire + V_{CC} drops across resistor R_2 driving point D and hence Y goes to ground i.e. 0 V or logic 0. i.e. A = B = 0, output Y = 0.

1.30 Transistor AND Gate

- As we have seen AND gate is logic gate in which output is high (logic 1) only when all the inputs are high (logic 1).

- AND gate can have two or more inputs, and even if one input is low (0), it makes output low (0).

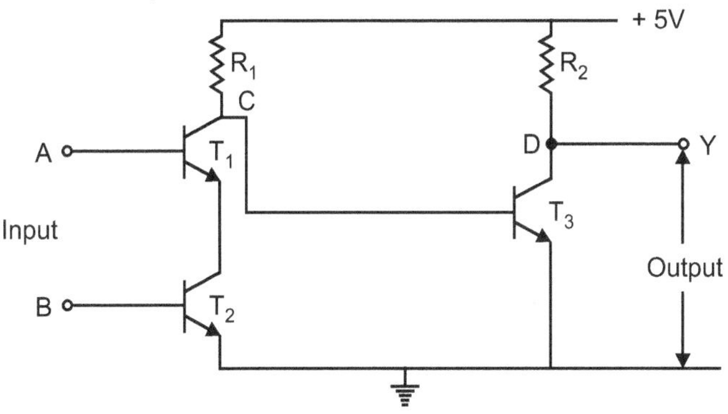

Fig. 1.15

- When either A = 0 V or B = 0 V or at logic 0, transistors T_1 and T_2 will be cut-off and there is no voltage drop across R_1. Hence, point C will go to supply voltage + 5 V and therefore transistor T_3 will conduct and whole supply voltage will be dropped across R_2. As a result, point D and hence output Y will go to 0 V or logic 0.

  i.e.   A = 0, B = 1 output Y = 0 and A = 1,   B = 0, output Y = 0.

- When both A and B are at + 5V or logic 1, the two transistors T_1 and T_2 conduct. The current so produced drops the supply voltage of

+ 5V across R_1, thereby driving point C and hence base of transistor T_3 to ground or 0 V. This will cut-off transistor T_3, so that output Y goes to supply voltage of + 5V or logic 1. i.e. A = 1   and   B = 1, output Y = 1.

1.31 Exclusive OR Gate (EX-OR Gate)

- The EX-OR operation is widely used in digital circuits. It is not a basic operation and can be performed using the basic gates AND, OR and NOT or universal gate NAND or NOR.

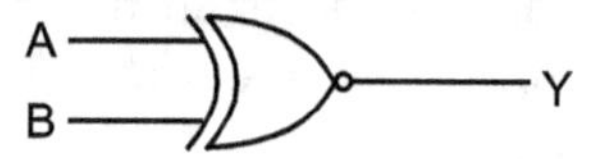

(a) Symbol

Inputs		Output
A	**B**	**Y = A ⊕ B**
0	0	0
0	1	1
1	0	1
1	1	0

(b) Truth table

Fig. 1.16

$$Y = \bar{A} B + A\bar{B}$$

i.e. $$Y = A \oplus B.$$

- The circle around plus ⊕ sign is worth noting. An EX-OR gate is special case of general OR gate. In general OR gate either input A or B or both must be 1 in order to get 1 at output. But in EX-OR case, the situation is different i.e. either input A or B, but not both should be 1 for getting 1 at the output.

- This circuit is also called an *inequality comparator* or *detector* because it produces an output only when the two inputs are different. EX-OR gate can be constructed using AND, OR and NOT gates.

- The upper AND gate forms the product $\bar{A}$ B while the lower AND gate produces $A\bar{B}$, the output of the OR gate is $Y = \bar{A} B + A\bar{B}$

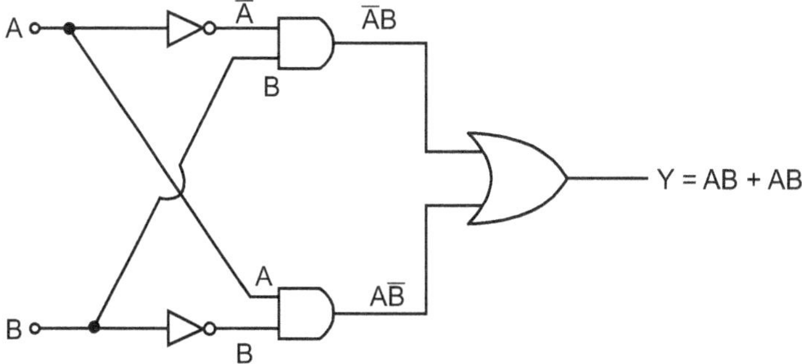

Fig. 1.17

Case I : If A and B both are LOW, both AND gates have low output, therefore, the final output Y is LOW.

i.e.   A = B = 0

$$\overline{A} \;=\; \overline{B} \;=\; 1$$

∴   Y = 1.0 + 0.1

                       = 0 + 0

                       = 0

Case II : If A = 0 (LOW) and B = 1 (HIGH), the upper AND gate has a HIGH output, so OR gate has HIGH output.

i.e.   A = 0   and   B = 1

$$\overline{A} \;=\; 1 \quad \text{and} \quad \overline{B} \;=\; 0$$

$$Y \;=\; \overline{A}\,B \;+\; A\overline{B}$$

                       = 1.1 + 0.0

                       = 1 + 0

                       = 1

Case III : Likewise   A = 1 (HIGH)   and B = 0 (LOW), Y output becomes 1 (HIGH).

Case IV : If A = 1 (HIGH) and B = 1 (HIGH), output Y becomes LOW.

i.e.   A = 1   and   B = 1

$$\overline{A} \;=\; 0 \quad \text{and} \quad \overline{B} \;=\; 0$$

                   Y = 0.1 + 1.0

                       = 0 + 0

                       = 0

1.32 Exclusive NOR Gate (EX-NOR Gate)

- Basically the "Exclusive-NOR" gate is a combination of the Exclusive-OR gate and the NOT gate but has a truth table similar to the standard NOR gate, in that, it has an output that is normally at logic level "1" and goes "LOW" to logic level "0" when ANY of its inputs are at logic level "1".

- However, an output "1" is only obtained if BOTH of its inputs are at the same logic level, either binary "1" or "0". For example, "00" or "11". This input combination would then give us the Boolean expression of : $Q = (A \oplus B) = A.B + \overline{A}.\overline{B}$.

1.33 Basics of Boolean Algebra and Karnaugh Map

- Boolean algebra is the mathematics of digital system. Infact it is a convenient and systematic way of expressing and analyzing the operation of logic circuits. It is found that the knowledge of Boolean algebra is essential to study and analysis of logic circuit.

- A mathematical system for formulating logical statements with symbols, so that problems can be written and solved like ordinary algebra, was developed by the Irish logician and mathematician George Boole and known as *Boolean algebra*.

- Boolean algebra was introduced in 1854. Boole is one of the persons in a long historical chain who were concerned with formalizing and mechanizing the process of logical thinking. Boolean algebra is a generalization of set algebra and the algebra of propositions and is a tool for studying and applying logic.

- It provides mathematical basis for expressing logic circuit functions, as well as analyzing and designing of the digital system.

1.34 Boolean Axioms

- The Boolean axioms are the laws of Boolean algebra for addition, multiplication and for the inversion. They are known as axioms because they are the truth which can be verified for different possibilities but cannot be proved.

- There are different operators for Boolean algebra.

Boolean Operators :

The operators in Boolean algebra are slightly different than conventional algebra for the basic gates and the following operators are commonly used.

(i)   Dot sign (·) : The dot sign (·) which is also expressed as (×), indicates logical product of two terms. The logical product of two terms A and B is expressed as A·B (or A × B) and is read as "A AND B". The inputs A and B are said to be ANDed.

(ii) Plus sign (+) : The (+) sign indicates the logical sum of two terms. For example, A + B represents logical sum of terms A and B and is read as "A OR B". The inputs A and B are said to be ORed.

(iii) Overbar (¯) : A sign of (¯) indicates that the terms which are overbar, are to be complemented. For example, $\overline{A}$ represents complementation of term A and is read as "NOT A".

Boolean Addition and Multiplication :

- Addition in Boolean algebra involves variables having values of either a binary 1 or a binary 0. Binary 1 will represent a HIGH level and binary 0 will represent a LOW level in Boolean equations.

    The basic rules for Boolean addition are as follows :

$$0 + 0 = 0$$
$$0 + 1 = 1$$
$$1 + 0 = 1$$
$$1 + 1 = 1$$

- Boolean addition is the same as the OR. Notice that, it differs from binary addition in the case where two 1s are added.

- Multiplication in Boolean algebra follows the same basic rules governing binary multiplication.

$$0 \cdot 0 = 0$$
$$0 \cdot 1 = 0$$
$$1 \cdot 0 = 0$$
$$1 \cdot 1 = 1$$

Boolean multiplication is the same as the AND.

Complementation in Boolean algebra is as follows :

$$\overline{0} = 1$$

$$\overline{1} = 0$$

It is same as NOT.

Rules and Laws of Boolean Algebra :

- There are certain rules and laws in Boolean algebra. The alphabets A, B, C and D can be used as variables having values 0 and 1.

Laws of Intersection :

Law 1 : $A \cdot 1 = A$

If a logic 1 is applied to one of the two inputs of the AND gate and signal A to the other input, the output will be A.

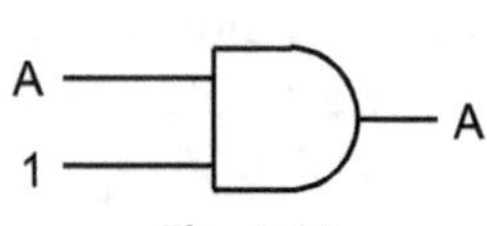
Fig. 1.18

Law 2 : $A \cdot 0 = 0$

If a logic 0 is applied to one of the two inputs of the AND gate and signal A to the other input, the output will be logic 0.

Fig. 1.19

Laws of Union :

Law 3 : $A + 1 = 1$

If a logic 1 is applied to the two inputs of OR gate and signal A to the other input, the output will be logic 1.

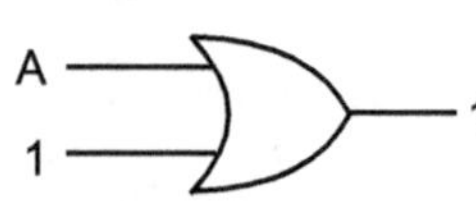
Fig. 1.20

Law 4 : $A + 0 = A$

If a logic 0 is applied to one of the two inputs of OR gate and A to the other input, the output will be logic A.

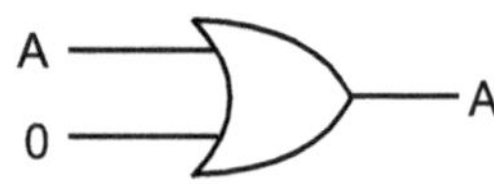
Fig. 1.21

Laws of Tautology :

Law 5 : $A \cdot A = A$

If the same signal A is applied to all the inputs of AND gate, the output will be same as the input.

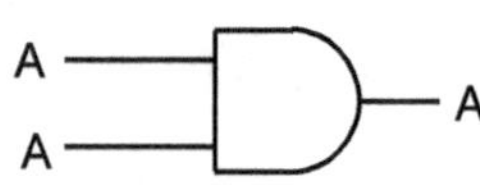
Fig. 1.22

Law 6 : $A + A = A$

If the same signal A is applied to all the inputs of OR gate, the output will be same as input A.

Fig. 1.23

Laws of Complement :

Law 7 : $A \cdot \overline{A} = 0$

If a logic signal A and its complement $\overline{A}$ is applied to an AND gate, the output will be logic 0.

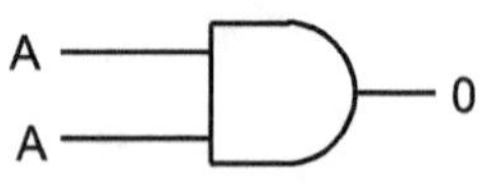
Fig. 1.24

Law 8 : $A + \bar{A} = 1$

If a logic signal A and its complement $\bar{A}$ is applied to an OR gate, the output will be 1.

Fig. 1.25

Law of Double Negation :

Law 9 : $\bar{\bar{A}} = A$

$$A \longrightarrow A \longrightarrow \bar{\bar{A}} = A$$

Fig. 1.26

The complement of the complement of A is A.

Laws of Commutation :

Law 10 : $A \cdot B = B \cdot A$

The commutative law of multiplication states that order in which variables are ANDed makes no difference at the output.

$$A,\ B \longrightarrow A \cdot B \quad \equiv \quad B,\ A \longrightarrow B \cdot A$$

Fig. 1.27

Law 11 : $A + B = B + A$

This law states that 'the order in which the inputs are given to an OR gate makes no difference at the output'.

$$A,\ B \longrightarrow A + B \quad \equiv \quad B,\ A \longrightarrow B + A$$

Fig. 1.28

Laws of Association :

Law 12 : The associative law of addition for three variables is stated as follows,

$$(A + B) + C \;=\; A + (B + C)$$

This law states that 'in the ORing of several variables, the result is same regardless of grouping of variables'.

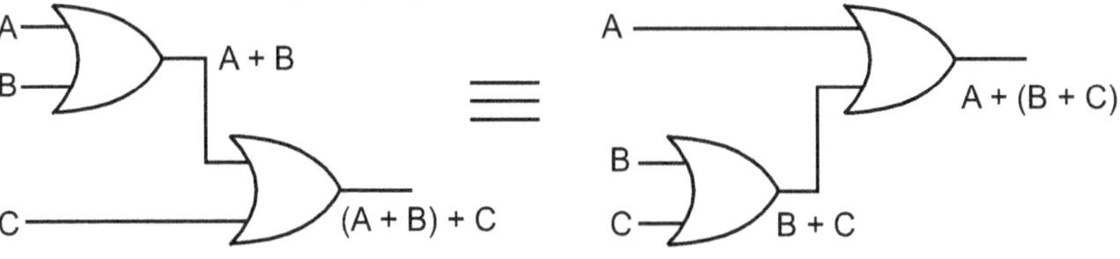

Fig. 1.29

Law 13 : The associative law of multiplication is stated as follows for three variables.

$$(A \cdot B) \cdot C \ = \ A \cdot (B \cdot C)$$

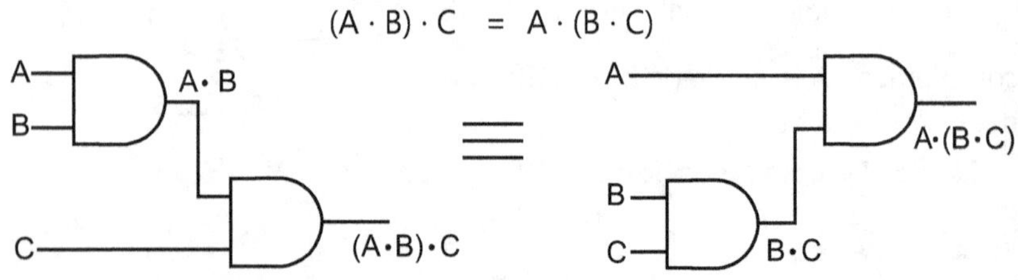

Fig. 1.30

Laws of Distribution:

Law 14 : The distributive law for three variables is written as follows :

$$A (B + C) \ = \ AB + AC$$

This law states that 'ORing several variables and ANDing the result with a single variable is equivalent to ANDing the single variable with each of several variables and then ORing the products'.

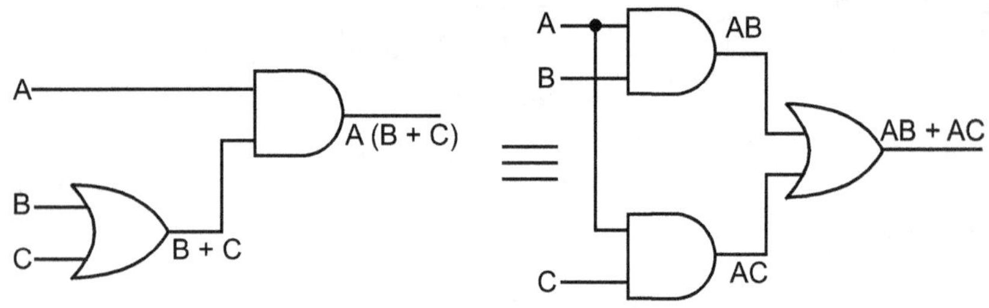

Fig. 1.31

Law 15 : $$A + (B \cdot C) \ = \ (A + B) \cdot (A + C)$$

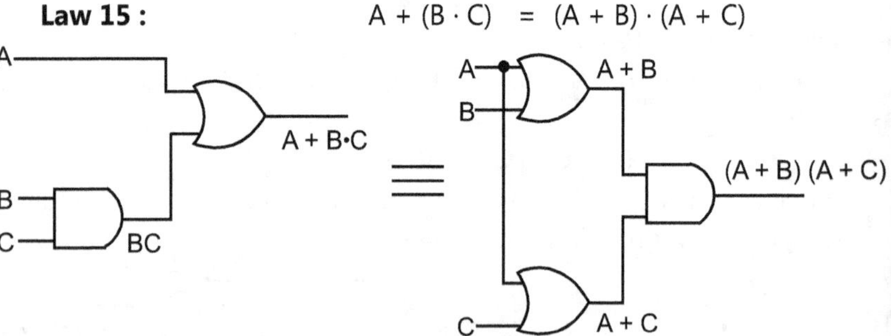

Fig. 1.32

Proof : $(A + B) \cdot (A + C) = AA + AB + AC + BC$

$$= \ A (1 + B + C) + BC \qquad \text{... (1)}$$

Though B and C have any value either 0 or 1,

$$1 + B + C = 1$$

Therefore equation (1) becomes $= A \cdot 1 + BC$

From law (1), $A \cdot 1 = A$

$$\therefore \qquad (A + B)(A + C) = A + BC$$

Laws of Absorption :

Law 16 : $\qquad A + A \cdot B = A$

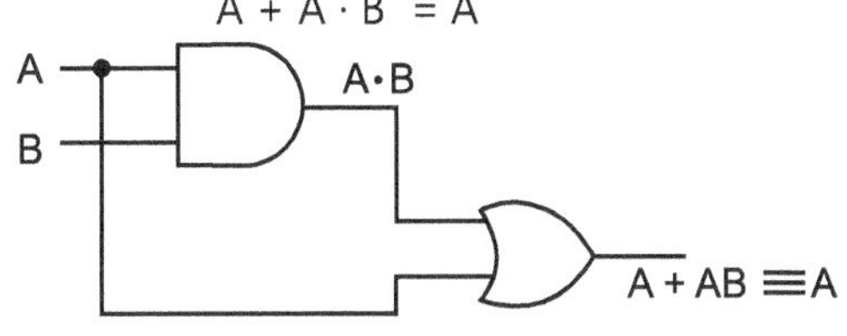

Fig. 1.33

Proof : Consider $\qquad A + A \cdot B = A \cdot (1 + B)$

$$= A \cdot 1 \qquad\qquad \because 1 + B = 1$$

$$= A \qquad\qquad \because A \cdot 1 = A$$

Law 17 : $\qquad A \cdot (A + B) = A$

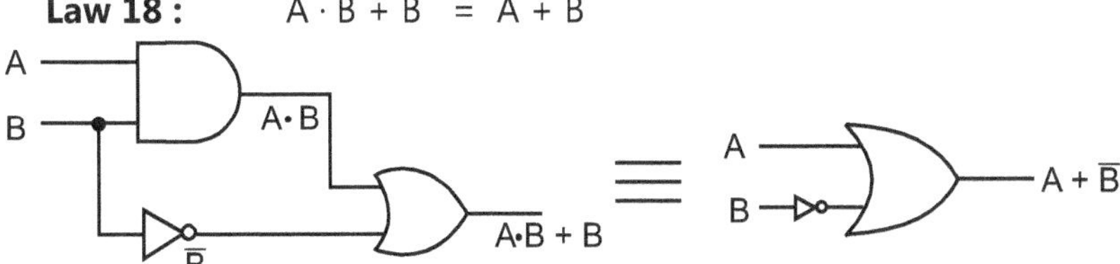

Fig. 1.34

Proof : $\qquad A \cdot (A + B) = A \cdot A + A \cdot B$

$$= A + AB \qquad\qquad \because A \cdot A = A$$

$$= A(1 + B)$$

$$= A \cdot 1 \qquad\qquad \because 1 + B = 1$$

$$= A \qquad\qquad \because A \cdot 1 = A$$

Law 18 : $\qquad A \cdot B + \bar{B} = A + \bar{B}$

Fig. 1.35

Proof :

$$A \cdot B + \bar{B} = A \cdot B + \bar{B} \cdot 1 \qquad \because \bar{B} \cdot 1 = \bar{B}$$

$$= A \cdot B + \bar{B} \cdot (A + 1) \qquad \because A + 1 = 1$$

$$= A \cdot B + \bar{B}A + \bar{B} \cdot 1$$

$$= A \cdot (B + \bar{B}) + \bar{B} \qquad \because \bar{B} \cdot 1 = \bar{B}$$

$$= A + \bar{B} \qquad \because B + \bar{B} = 1$$

Law 19 : $\quad A \cdot (\bar{A} + B) = A \cdot B$

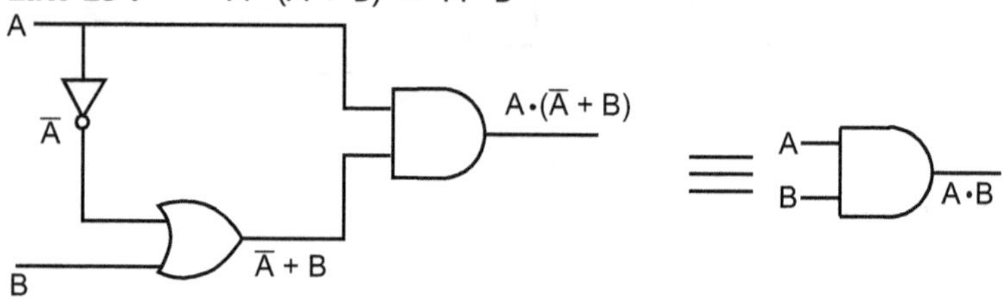

Fig. 1.36

Proof : $\quad A \cdot (\bar{A} + B) = A\bar{A} + AB$

$$= 0 + AB \qquad \because A \cdot \bar{A} = 0$$

$$= A \cdot B$$

Law 20 : $\quad A \cdot \bar{B} + B = A + B$

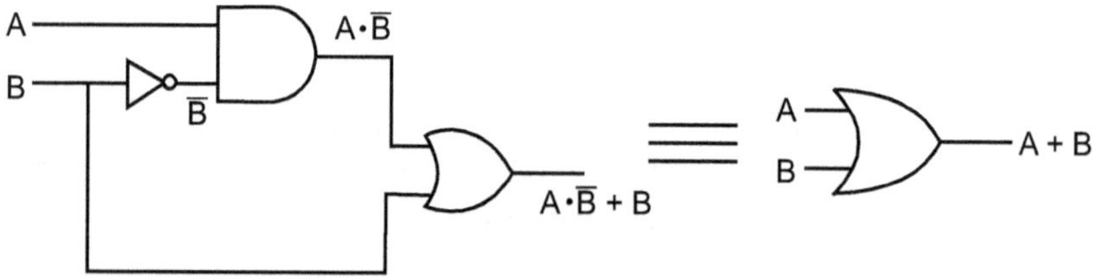

Fig. 1.37

Proof : Consider $A \cdot \bar{B} + B = A \cdot \bar{B} + B \cdot 1 \qquad \because B \cdot 1 = B$

$$= A \cdot \bar{B} + B \cdot (A + 1) \qquad \because A + 1 = 1$$

$$= A \cdot \bar{B} + B \cdot A + B \cdot 1$$

$$= A (\bar{B} + B) + B$$

$$= A \cdot 1 + B \qquad \because \bar{B} + B = 1$$

$$= A + B \qquad \because A \cdot 1 = A$$

Table 1.11 : Boolean algebraic theorems

Law	Classification	Algebraic definition
1	Laws of intersection	$A \cdot 1 = A$
2		$A \cdot 0 = 0$
3	Laws of union	$A + 1 = 1$
4		$A + 0 = A$
5	Laws of tautology	$A \cdot A = A$
6		$A + A = A$
7	Laws of complement	$A \cdot \bar{A} = 0$
8		$A + \bar{A} = 1$
9	Law of double negation	$\bar{\bar{A}} = A$
10	Laws of commutation	$A \cdot B = B \cdot A$
11		$A + B = B + A$
12	Laws of association	$(A + B) + C = A + (B + C)$
13		$(A \cdot B) \cdot C = A \cdot (B \cdot C)$
14	Laws of distribution	$A (B + C) = A \cdot B + A \cdot C$
15		$A + (B \cdot C) = (A + B) \cdot (A + C)$
16	Laws of absorption	$A + A \cdot B = A$
17		$A \cdot (A + B) = A$
18		$A \cdot B + \bar{B} = A + \bar{B}$
19		$A \cdot (\bar{A} + B) = AB$
20		$A \cdot \bar{B} + B = A + B$

1.35 Simplification of Logic Equations Using Laws of Boolean Algebra

- Many times it is essential to reduce number of gates required for designing a digital circuit, reduce a particular expression to its simplest form, ultimately which reduces size and price (cost) of circuit.

- By applying basic laws, rules and theorems of Boolean algebra, it is possible to implement practically.

Following examples illustrate the technique.

Example 1 : Simplify the expression AB + A (B + C) + B (B + C) using Boolean algebra techniques.

Solution : Consider

$$
\begin{aligned}
AB + A (B + C) + B (B + C) &= AB + AB + AC + BB + BC \\
&= A (B + B) + AC + B (B + C) \quad \because B \cdot B = 1 \\
&= AB + AC + B \quad \because B + B = B,\ B (B + C) = B \\
&= AB + B + AC \\
&= B (A + 1) + AC \\
&= B \cdot 1 + AC \qquad\qquad \because A + 1 = 1 \\
&= B + AC
\end{aligned}
$$

$$
\therefore \qquad AB + A (B + C) = B + AC
$$

The logic circuit will be

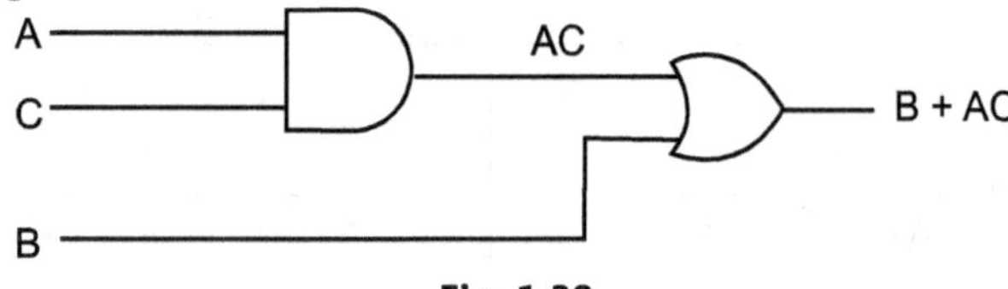

Fig. 1.38

Example 2 : Simplify the expression [A$\overline{\text{B}}$ (C + BD) + $\overline{\text{A}}$ $\overline{\text{B}}$] C using Boolean algebra techniques.

Solution : Consider

$$
\begin{aligned}
[A\overline{B} (C + BD) + \overline{A}\,\overline{B}] C &= (A\overline{B}C + A\overline{B}BD + \overline{A}\,\overline{B}) C \\
&= (A\overline{B}C + A \cdot 0 \cdot D + \overline{A}\,\overline{B}) C \qquad \because \overline{B} B = 0 \\
&= (A\overline{B}C + 0 + \overline{A}\,\overline{B}) C \\
&= (A\overline{B}C + \overline{A}\,\overline{B}) C \\
&= A\overline{B}CC + \overline{A}\,\overline{B}C \\
&= A\overline{B}C + \overline{A}\,\overline{B}C \qquad\qquad \because C \cdot C = C \\
&= \overline{B}C (A + \overline{A}) \\
&= \overline{B}C \cdot (1) \qquad\qquad \because A + \overline{A} = 1 \\
&= \overline{B}C
\end{aligned}
$$

$$
\therefore \qquad [A\overline{B} (C + BD) + \overline{A}\overline{B}] C = \overline{B}C
$$

The logic circuit will be

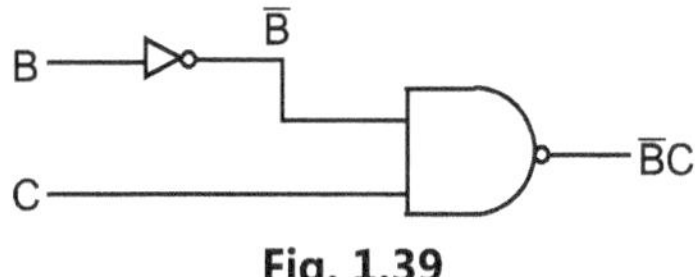

Fig. 1.39

Example 3 : Using Boolean algebra techniques, simplify the following expressions as much as possible : (i) A(A + B), (ii) A($\overline{A}$ + AB), (iii) BC + $\overline{B}$C, (iv) A(A + $\overline{A}$B).

Solution : (i)

$$A (A + B) = A{\cdot}A + A{\cdot}B$$
$$= A + A{\cdot}B \qquad \because A{\cdot}A = A$$
$$= A (1 + B)$$
$$= A \cdot (1) \qquad \because 1 + B = 1$$
$$= A$$
$$\therefore \quad A (A + B) = A$$

(ii)

$$A (\overline{A} + AB) = A\overline{A} + A{\cdot}AB$$
$$= 0 + A{\cdot}AB \qquad \because A\overline{A} = 0$$
$$= AB \qquad \because A{\cdot}A = A$$
$$\therefore \quad A (\overline{A} + AB) = AB$$

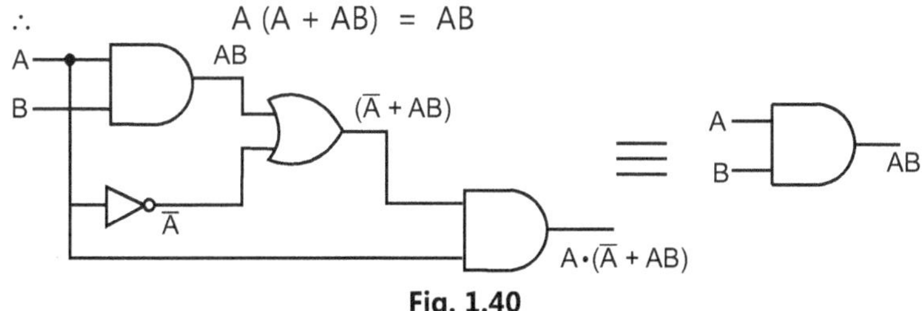

Fig. 1.40

(iii)

$$BC + \overline{B}C = C (B + \overline{B})$$
$$= C \cdot (1) \qquad \because B + \overline{B} = 1$$
$$= C$$

(iv)

$$A (A + \overline{A}B) = A{\cdot}A + A{\cdot}\overline{A}B$$
$$= A + A{\cdot}\overline{A}B \qquad \because A{\cdot}A = A$$
$$= A + 0 \qquad \because A{\cdot}\overline{A} = 0$$
$$= A \qquad \because A + 0 = A$$

Example 4 : Reduce the following Boolean expression and draw the logic diagram :

$$\bar{A}BC\bar{D} + BC\bar{D} + \bar{B}\bar{C}\bar{D} + B\bar{C}D$$

Solution : Consider

$$\bar{A}BC\bar{D} + BC\bar{D} + B\bar{C}\bar{D} + B\bar{C}\,D \;=\; BC\bar{D}\,(\bar{A} + 1) + B\bar{C}\,(\bar{D} + D)$$

$$(\because \bar{A} + 1 = 1,\; D + \bar{D} = 1)$$

$$= BC\bar{D}\cdot(1) + B\bar{C}\cdot(1) = BC\bar{D} + B\bar{C}$$

$$= B\,(C\bar{D} + \bar{C}\,)$$

$$= B\,(\bar{D}C + \bar{C}\,) \qquad\qquad (\because \bar{D}C = C\bar{D})$$

$$= B\,(\bar{D} + \bar{C}\,) \qquad (\because \bar{D}C + \bar{C} = \bar{D} + \bar{C}\,)$$

The logic circuit will be

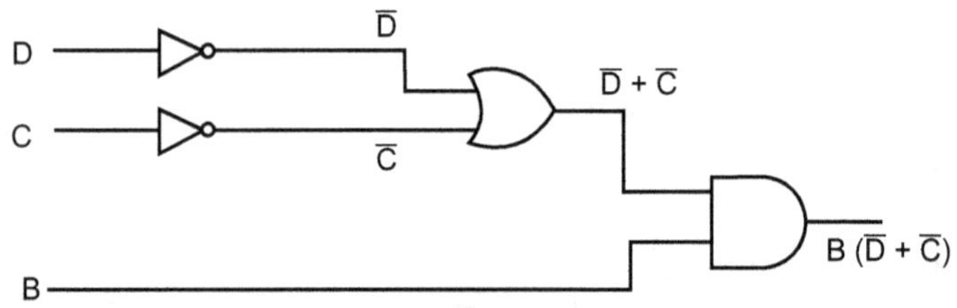

Fig. 1.41

Example 5 : Simplify the following equations using laws of Boolean algebra : (i) $Y = ABCD + ABC + AB + A\bar{B}$

Solution : Consider

$$ABCD + ABC + AB + A\bar{B} \;=\; ABC\,(D + 1) + A\,(B + \bar{B})$$

$$= ABC\cdot(1) + A\cdot(1)$$

$$(\because D + 1 = 1,\; B + \bar{B} = 1)$$

$$= ABC + A$$

Circuit for $Y = ABCD + ABC + AB + A\bar{B}$ can be constructed as follows :

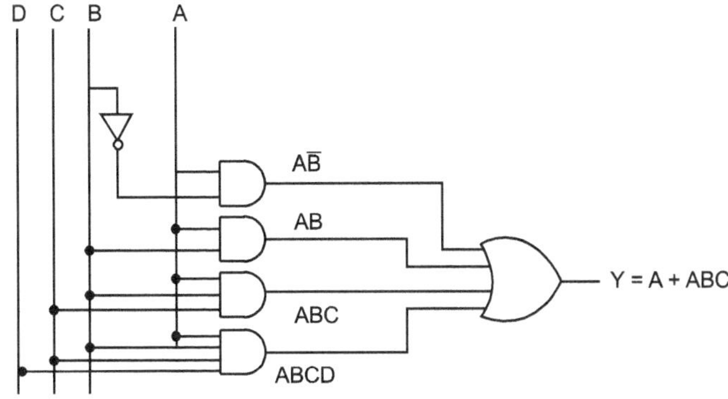

Fig. 1.42

Here four input AND and four input OR gate are used.

(ii) $Y = \bar{A} + AB + A\bar{B}$

Solution : Consider

$$\bar{A} + AB + A\bar{B} = \bar{A} + A (B + \bar{B})$$

$$= \bar{A} + A (1) = \bar{A} + A = 1 \qquad (\because B + \bar{B} = 1)$$

Fig. 1.43

Here three input OR gate is used.

(iii) $Y = AB + \bar{A}B + ABC$

Solution : $Y = AB + ABC + \bar{A}B$

$$= AB (1 + C) + \bar{A}B$$

$$= AB (1) + \bar{A} B = AB + \bar{A}B \qquad (\because 1 + C = 1)$$

$$= B (A + \bar{A})$$

$$= B \qquad (\because A + \bar{A} = 1)$$

Let us consider the following circuit to solve the given equation.

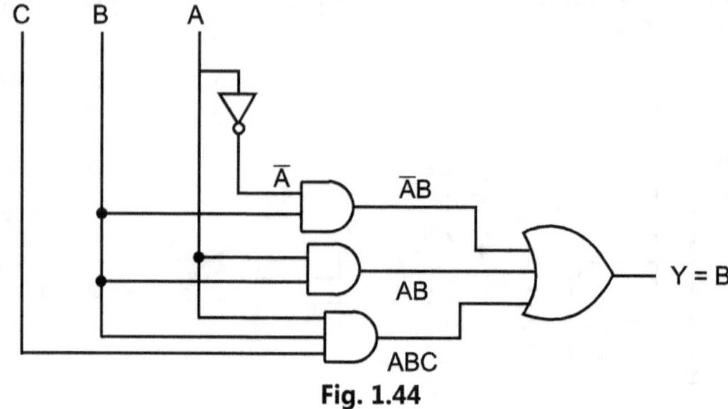

Fig. 1.44

(iv) $A = AB + BC + \bar{B}A + \bar{A}B$

Solution :

$$Y = AB + \bar{A}B + BC + \bar{B}A = B(A + \bar{A}) + BC + \bar{B}A$$

$$= B + BC + \bar{B}A \qquad\qquad (\because A + \bar{A} = 1)$$

$$= B(1 + C) + \bar{B}A$$

$$= B + \bar{B}A \qquad\qquad (\because 1 + C = 1)$$

Let us consider the following circuit to solve the given equation.

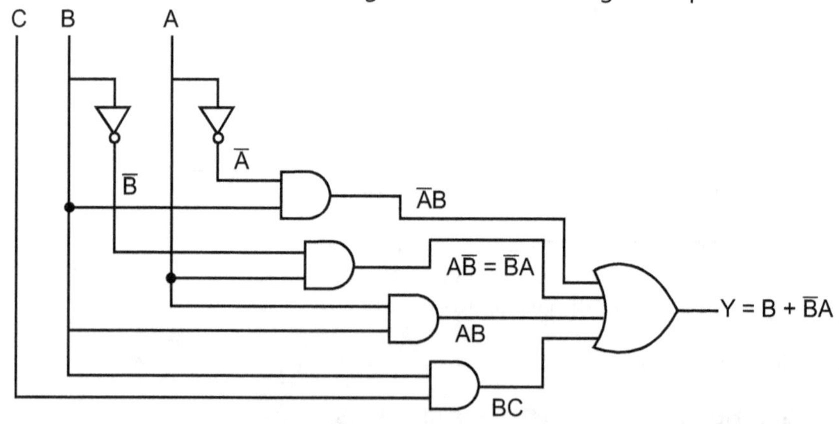

Fig. 1.45

1.36 Boolean Expression in SOP and POS Form

- Boolean expressions can be used to build the logic circuit. If we have the expression $Y = A + B + C$ and asked to build a circuit that perform this logic function, we can very easily see that there must be an OR gate having three inputs A, B and C.

- The circuit can be realised using three input OR gate as shown in Fig. 1.46.

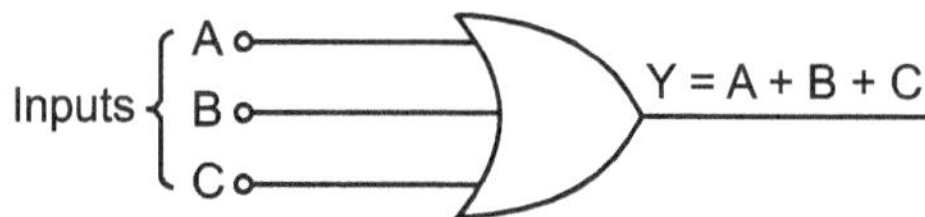

Fig. 1.46 : Logic diagram for Boolean expression Y = A + B + C

- It is found that, the Boolean expression come in two forms. All Boolean expressions can be converted into either of two standard forms; the sum of product form or the product of sum forms.

- It is found that this standardization makes the evaluation, simplification and implementation of Boolean expressions much more systematic and easier.

Sum of Product (SOP) Form :

- A product term consist of the product of multiplication of variables or their complements (known as **literal**).

- When two or more product terms are summed by Boolean addition, the resulting expression is known as sum of product (SOP) form. The examples of SOP are,

$$Y = AC + BC$$

$$Y = AB + ABC$$

$$Y = ABC + BCD + AB\overline{D} \text{ etc.}$$

- The sum of product form is called **minterm** *form* in engineering texts and the product of sum form is called the **maximum** *term form* by engineers, technicians and scientist.

- An SOP expression can contain a single variable term. In an SOP expression a single overbar cannot extend over more than one variable, however more than one variable in a term can have an overbar. Thus, $\overline{ABC}$ is not valid but $\overline{A}\,\overline{B}\,\overline{C}$ is valid in SOP.

- The set of variables contained in the expression in either complemented or uncomplemented form is known as domain of a Boolean expression. For example, the domain of the expression $A\overline{B}$ + ABC is the set of variable A, B and C.

Implementation of an SOP Expression:

- The sum of product (SOP) expression contains product of sum terms. A product term is produced by an AND operation and the sum or addition of two or more product terms is produced by an OR operation.

- Therefore, an SOP expression is implemented by an AND-OR logic. Consider the SOP expression y = AB + BC + CD, it can be implemented as shown in Fig. 1.47.

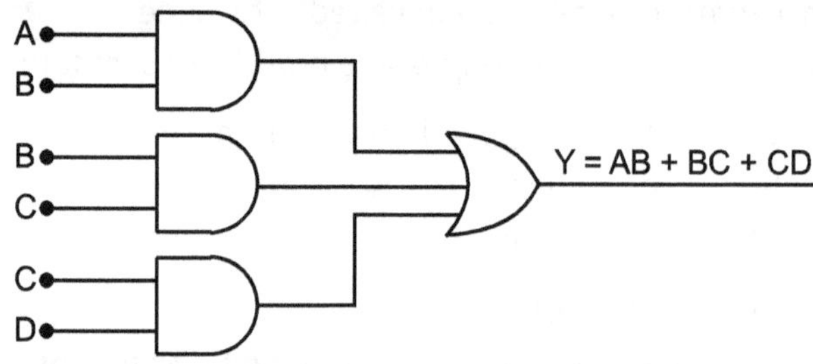

Fig. 1.47 : SOP of Boolean expression Y = AB + BC + CD

The Product of Sum (POS) Form :

- A sum term consist of the sum i.e. Boolean addition of literals (variables or their complements). When two or more sum terms are multiplied, the resulting expression is known as **product of sum (POS)** form, for example,

$$Y = (A + B)(B + C)$$

$$Y = (\overline{A} + B)(A + \overline{B} + C)$$

- POS expression can contain a single variable term. In a POS expression, a single overbar cannot extend over more than one variable, however more than one variable in a term can have an overbar. For example, POS expression can have the term $\overline{A} + \overline{B} + \overline{C}$ but will not have $\overline{A + B + C}$.

Implementation of a POS Expression :

- Given Boolean expression, of the POS form can be implemented by ANDing the outputs of two or more OR gates.

- A sum term is produced by an OR operation and the product of two or more sum terms is produced by an AND operation.

- The expression Y = (A + B) (B + C) (C + D) can be implemented as shown in Fig. 1.48 POS form.

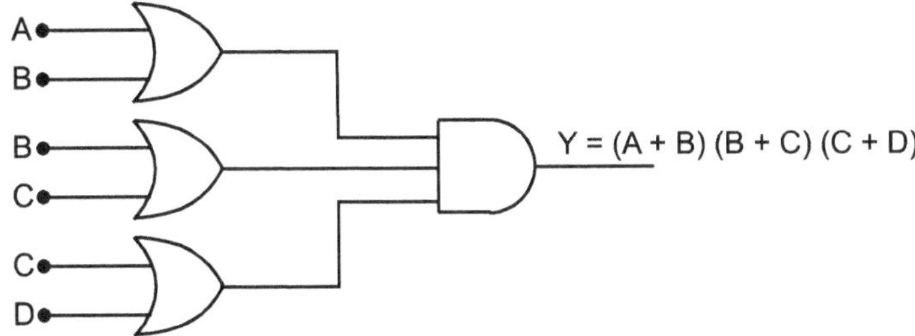

Fig. 1.48 : POS form of Y = (A + B) (B + C) (C + D)

Standard or Canonical SOP and POS forms :

- A logic expression is said to be in the standard or canonical form. If each SOP term consists of all the literals/variables in their complemented or uncomplemented form, the standard SOP form is

$$Y = ABC + A\overline{B}C + AB\overline{C}$$

Each product term
contain all the three variables

The Boolean expression is said to be in standard POS form, if all the terms in POS consist of all the literals in their complemented or uncomplemented form. For example,

$$Y = (A + B + C) \cdot (\overline{A} + \overline{B} + \overline{C}) \cdot (A + B + \overline{C})$$

Each sub term
contain all the literals

Each individual term in the standard SOP form is called *minterm*.

e.g. Consider,

$$Y = ABC + AB\overline{C} + \overline{A}B\overline{C}$$

Each individual terms
called minterm

Each individual term in the standard POS form is called *maxterm*.

e.g. Consider,

$$Y = (A + B) + (\overline{C} + \overline{B})$$

Each individual terms
called maxterm

1.37 Conversion of SOP/POS Expression to Its Standard SOP/POS Form

- The given Boolean expression can be converted to their corresponding SOP and POS forms. As seen in the standard SOP form each product term consist of all the literals.

e.g. $Y = AB + A\bar{B} + \bar{A}\bar{B}$ is standard SOP

but $Y = AB + ABC + AB\bar{C}$ is not standard SOP

- The conversion of expression into standard SOP form is a three step process :

(1)   For each term find the missing literal,

(2)   Then AND the term with the term formed by ORing the missing literal and its complement,

(3)   Simplify the obtained equation.

e.g. Convert the expression

$$Y = AB + A\bar{C} + BC \text{ in the standard SOP form.}$$

Solution : Given expression is

$$Y = AB + A\bar{C} + BC$$

Step 1 : Find the missing literal in each term.

$$Y = AB \quad + \quad A\bar{C} \quad + \quad BC$$

| Missing literal is C | Missing literal is B | Missing literal is A |

Step 2 : AND each term with its (Missing literal + Its complement).

$$Y = AB(C + \bar{C}) + A\bar{C}(B + \bar{B}) + BC(A + \bar{A})$$

Note that $C + \bar{C} = 1$ then it does not changes the value of expression.

Step 3 : Simplification of the expression.

$$Y = AB(C + \bar{C}) + A\bar{C}(B + \bar{B}) + BC(A + \bar{A})$$

$$Y = ABC + AB\bar{C} + AB\bar{C} + A\bar{B}\bar{C} + ABC + \bar{A}BC$$

$$= (ABC + ABC) + (AB\bar{C} + AB\bar{C}) + A\bar{B}\bar{C} + \bar{A}BC$$

Since $A + A = A$

$$Y = ABC + AB\bar{C} + A\bar{B}\bar{C} + \bar{A}BC$$

Since in the above expression each term contain all the literals it is in the standard SOP form.

Conversion to Standard POS Form

* In the standard POS form each sum term consists of all the literals in the complemented or uncomplemented form.

e.g. $Y = (\bar{A} + B) \cdot (\bar{A} + \bar{B}) \cdot (A + \bar{B})$ is in the standard POS

whereas, $Y = (\bar{A} + \bar{B}) \cdot (A + B + C)$ is in the non-standard POS form.

The given POS expression can be converted into standard POS form using three steps.

Step 1 : For each term find the missing literal.

Step 2 : Then OR each term with the term formed by ANDing the missing literal in that term with its complement.

Step 3 : Simplify the expression to obtain standard POS form.

e.g. Convert the expression

$$Y = (A + B) \cdot (A + \bar{B}) \cdot (B + \bar{C}) \text{ into the standard POS forms.}$$

Solution : Find the missing literal in each term

$$Y = (A + B) \cdot (A + \bar{B}) \cdot (B + \bar{C})$$

Missing literal is C ; Missing literal is C ; Missing literal is A

Step 1 : OR each term with missing literal and its complement.

$$Y = (A + B + C\bar{C}) \cdot (A + \bar{B} + C\bar{C}) \cdot (B + \bar{C} + A\bar{A})$$

Step 2 : Simplify the expression

$$Y = (A + B + C\bar{C}) \cdot (A + \bar{B} + C\bar{C}) \cdot (B + \bar{C} + A\bar{A}) \qquad \text{... (1.1)}$$

Let $p = A + B, \quad q = A + \bar{B}, \quad r = B + \bar{C},$

$\therefore$ $Y = (p + C\bar{C}) \cdot (q + C\bar{C}) \cdot (r + A\bar{A}) \qquad \text{... (1.2)}$

Since, $A + BC = (A + B)(A + C)$

$$= (A + B + C)(A + B + \bar{C})(A + \bar{B} + C)(A + \bar{B} + \bar{C})$$

$$(B + \bar{C} + A)(B + \bar{C} + \bar{A})$$

$$\therefore \quad Y = (A + B + C)(A + B + \bar{C})$$

$$(A + \bar{B} + C)(A + \bar{B} + \bar{C})(\bar{A} + B + \bar{C})$$

Since each term of above expression contains all the literals so the equation is standard POS form.

1.38 De Morgan's Theorems

1.38.1 De Morgan's First Theorem

De Morgan, a logician and mathematician proposed two theorems which are important parts of Boolean algebra.

De Morgan's first theorem is,

$$\overline{AB} = \bar{A} + \bar{B}$$

The complement of product is equal to the sum of the complements. The complement of two or more variables ANDed is the same as the OR of the complement of each individual variables.

Fig. 1.49

It can also be constructed with AND, OR and NOT gate as follows :

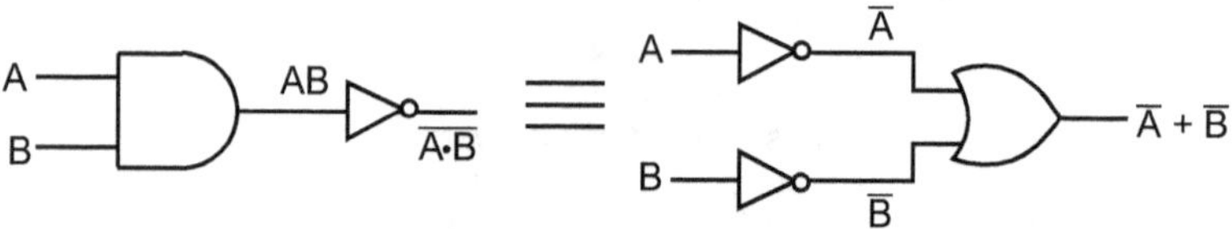

Fig. 1.50

Table 1.12 : De Morgan's first theorem

Input		Intermediate value			Output	
A	**B**	**AB**	**$\bar{A}$**	**$\bar{B}$**	**$\overline{AB}$**	**$\bar{A} + \bar{B}$**
0	0	0	1	1	1	1
0	1	0	1	0	1	1
1	0	0	0	1	1	1
1	1	1	0	0	0	0

1.38.2 De Morgan's Second Theorem

The complement of a sum is equal to the product of the complements. It can be expressed as

$$\overline{A + B} = \overline{A} \cdot \overline{B}$$

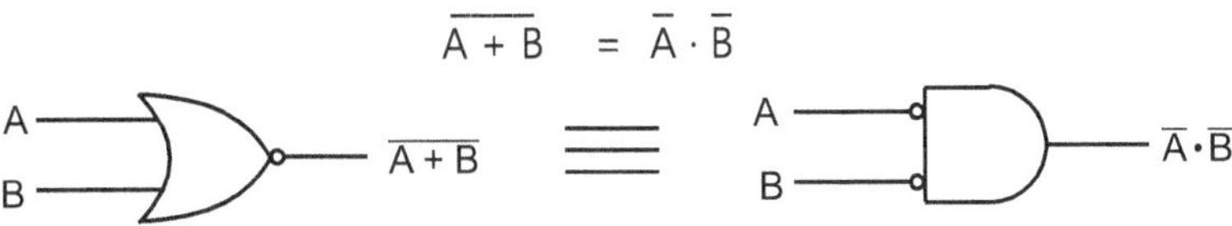

Fig. 1.51

Using AND, OR and NOT gate, the circuit can be drawn as

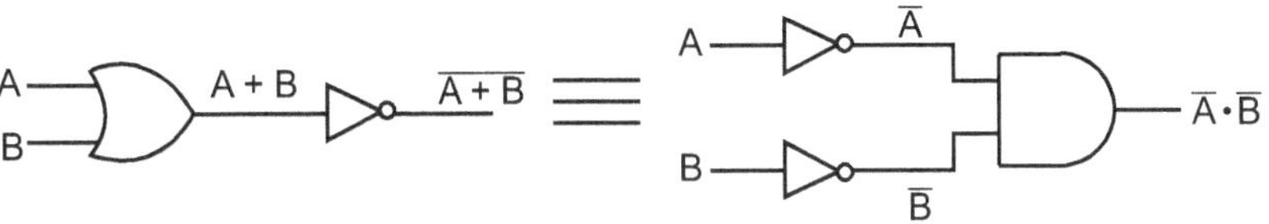

Fig. 1.52

Table 1.13 : De Morgan's second theorem

Input		Intermediate value			Output	
A	B	A + B	$\overline{A}$	$\overline{B}$	$\overline{A + B}$	$\overline{A} \cdot \overline{B}$
0	0	0	1	1	1	1
0	1	1	1	0	0	0
1	0	1	0	1	0	0
1	1	1	0	0	0	0

Now, consider the NAND operation of three variables.

$$\overline{ABC} = \overline{A} + \overline{B} + \overline{C}$$

and
$$\overline{A + B + C} = \overline{A} \cdot \overline{B} \cdot \overline{C}$$

The above results can be easily extended to any number of variables.

1.38.3 Applications of De Morgan's Theorems

De Morgan's theorems can be used for simplifying logic function.

Example 1 : Simplify the following expressions.

Solution : (i)
$$\overline{\overline{A + B} + \overline{C}} = \overline{\overline{A + B}} \cdot \overline{\overline{C}}$$

$$[\because \overline{\overline{A + B}} = A + B \text{ and } \overline{\overline{C}} = C]$$

$$= \overline{A + B} \cdot C$$
$$= (A + B) \cdot C$$

(ii) $\overline{\overline{A+B}+\overline{CD}} = \overline{\overline{A+B}}\cdot\overline{\overline{CD}} = (A+B)\,CD$

(iii) $\overline{\overline{AB}+\overline{A}+AB} = \overline{\overline{A+B}+\overline{A}+AB}$

$$= \overline{\overline{A}+\overline{B}+AB} \qquad \because \overline{A}+\overline{A}=\overline{A}$$

$$= \overline{\overline{A}+A+\overline{B}} \qquad \because AB+\overline{B}=A+\overline{B}$$

$$= \overline{\overline{1+\overline{B}}} = \overline{\overline{1}} = 0$$

$$\overline{\overline{AB}+\overline{A}+AB} = 0$$

(iv) $\overline{(\overline{A}\cdot B)\,(B\cdot C)\,(C\cdot\overline{D})} = \overline{\overline{A}\cdot B} + \overline{B\cdot C} + \overline{(C\cdot\overline{D})}$

$$= \overline{\overline{A}}+\overline{B}+\overline{B}+\overline{C}+\overline{C}+\overline{\overline{D}}$$

$$= A+\overline{B}+\overline{B}+\overline{C}+\overline{C}+D$$

$$\qquad \because \overline{\overline{A}}=A \text{ and } \overline{\overline{D}}=D$$

$$= A+\overline{B}+\overline{C}+D$$

$$\qquad \because \overline{B}+\overline{B}=\overline{B} \text{ and } \overline{C}+\overline{C}=\overline{C}$$

$\therefore$ $\overline{(\overline{A}\cdot B)\,(B\cdot C)\,(C\cdot\overline{D})} = A+\overline{B}+\overline{C}+D$

(v) $y = \overline{(\overline{A}\cdot B)+(A\cdot\overline{B})}$

Using De Morgan's theorem

$$\overline{A+B} = \overline{A}\,\overline{B}$$

$$y = \overline{(\overline{A}\cdot B)}\cdot\overline{(A\cdot\overline{B})}$$

$\therefore$ $y = (\overline{A}\cdot B)\cdot(A\cdot\overline{B}) \qquad (\because \overline{\overline{A}}=A)$

1.39 Universal Gate (NAND/NOR): Multilevel NAND and NOR Gates

Table 1.14

Function	Symbol	NAND	NOR
1. NOT	$A \longrightarrow \overline{A}$	$A \longrightarrow \overline{A}$	$A \longrightarrow \overline{A}$
2. AND	$A, B \longrightarrow AB$	$A, B \longrightarrow \overline{A \cdot B} \longrightarrow A \cdot B$	$A, B \longrightarrow A \cdot B$
3. OR	$A, B \longrightarrow A + B$	$A, B \longrightarrow A + B$	$A, B \longrightarrow \overline{A + B} \longrightarrow A + B$

The NAND and NOR gates have universal property and can be used for performing AND, OR and NOT functions, so an AND/OR/NOT logic circuit can be converted to NAND/NOR logic.

1.39.1 NAND as NOT Gate

$$Y = \overline{A \cdot A} = \overline{A}$$

A NOT gate can be obtained from NAND gate by connecting all the inputs together.

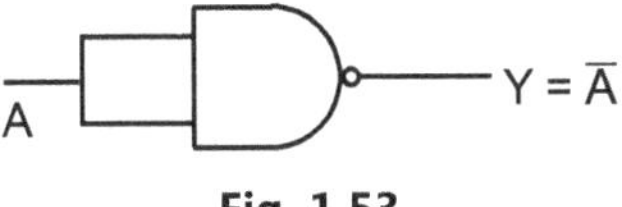

Fig. 1.53

1.39.2 NAND as AND Gate

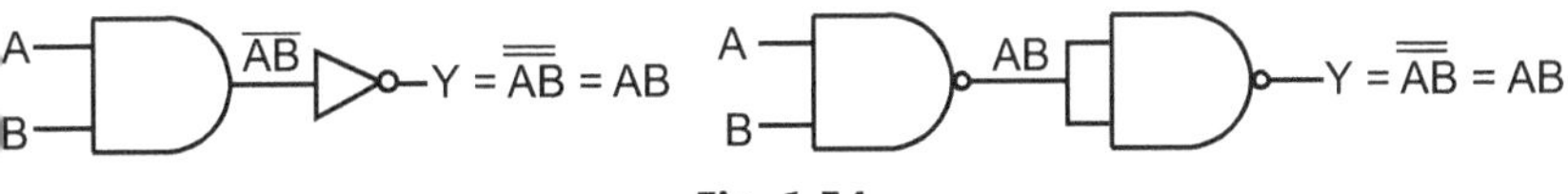

Fig. 1.54

The AND gate can be obtained by simply inverting output of NAND gate.

1.39.3 NAND as OR Gate

The OR gate can be designed by using NAND gate.

$$Y = A + B$$

$$Y = \overline{\overline{A}} + \overline{\overline{B}}$$

$$Y = \overline{\overline{A} \cdot \overline{B}}$$

Therefore an OR gate operation can be obtained by NANDing $\bar{A}$ & $\bar{B}$.

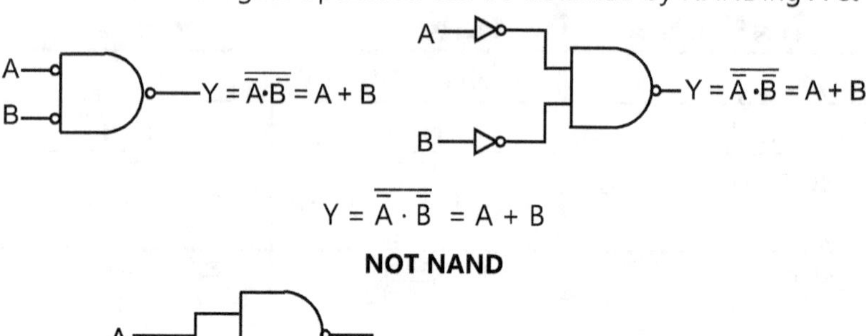

$$Y = \overline{\overline{A} \cdot \overline{B}} = A + B$$

NOT NAND

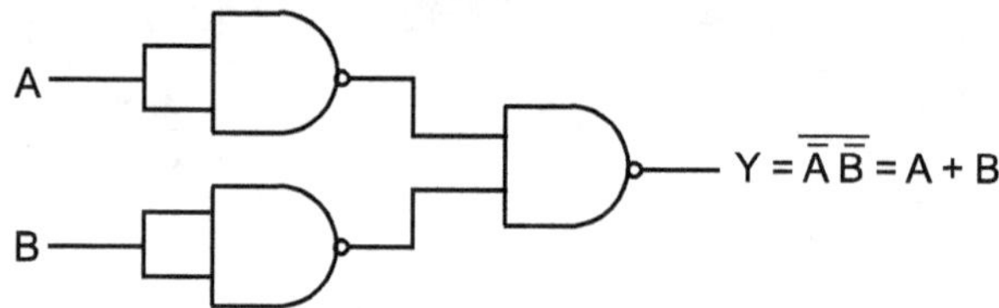

$$Y = \overline{\overline{A}} \ \overline{\overline{B}} = A + B$$

Fig. 1.55 : OR gate using NAND gate

1.39.4 NOR as NOT Gate

$$A \qquad Y = \overline{A + A} = \overline{A}$$

Fig. 1.56

Two inputs of OR gate are connected together, it gives inverter of input signal i.e. NOT.

1.39.5 NOR as AND Gate

The AND gate can be designed by using NOR gate.

$$Y = A \cdot B$$

$$Y = \overline{\overline{A}} \cdot \overline{\overline{B}}$$

$$Y = \overline{\overline{A} + \overline{B}}$$

$$Y = \overline{\overline{A} + \overline{B}} = A \cdot B \qquad\qquad Y = \overline{\overline{A} + \overline{B}} = A \cdot B$$

$$Y = \overline{\overline{A} + \overline{B}} = A \cdot B$$

Fig. 1.57

1.40 Karnaugh Map

- The Karnaugh map or K-map provides a systematic method for simplifying a Boolean expression and can be used as visual display of fundamental products needed for a sum of products solution.

- The K-map is composed of an arrangement of adjacent 'cells' each representing one particular combination of variables in product form. The K-map consists of 2^n cells, where n is number of variables.

- For example, there are four combinations of the products of two variables A and B and their complements $\overline{A}\,\overline{B}$, $\overline{A}\,B$, $A\overline{B}$ and AB. Therefore, the K-map must have four cells, with each cell representing one of the variable combination.

1.40.1 Advantages of K-map

- As we have seen the laws of Boolean algebra, but it is difficult to apply these laws when number of variables are large. In such a case, it makes easy by using K-map.

- Use of large number of laws of Boolean algebra increases the chances of error because one have to remember all these laws and has to apply them at correct place. But by using K-map it is not necessary to remember all the laws, according to circuit situation K-map may apply.

- It provides easiest way to produce simplest Boolean expression with minimum chances of error.

  Format of a two variable K-map can be represented as follows:

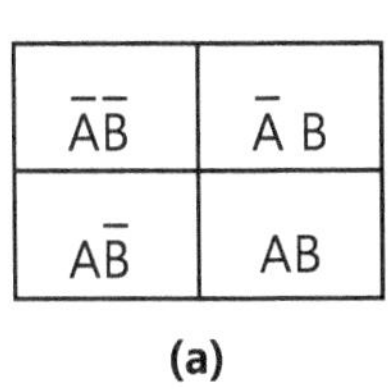

(a)

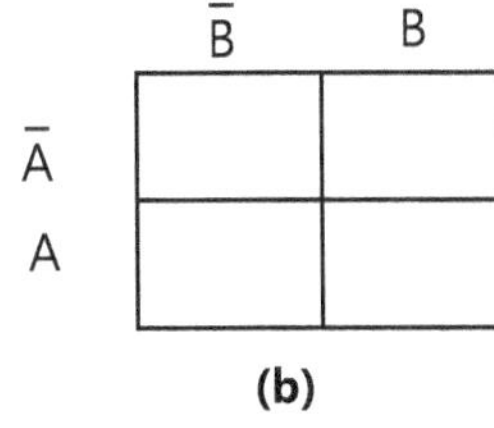

(b)

Fig. 1.58

- Variable combination is shown in Fig. 1.58 (a) and actually how K-map can be arranged with variables outside the cell is shown in Fig. 1.58 (b).

Extensions of the K-map to three and four variables are shown in Fig. 1.59.

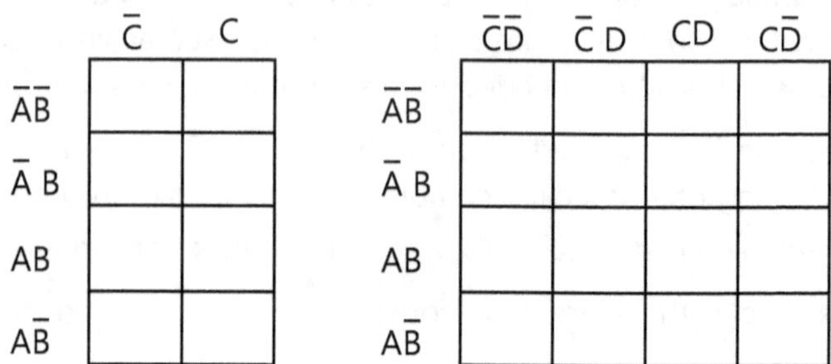

(a) Three variable map (2^3 = 8 cells)   (b) Four variable map (2^4 = 16 cells)

Fig. 1.59

Karnaugh maps can be used for five, six or more variables.

We see how K-map can be drawn from the following Table 1.14. Consider the Table 1.15.

Table 1.15

A	B	C	Y
0	0	0	0
0	0	1	0
0	1	0	1
0	1	1	0
1	0	0	0
1	0	1	0
1	1	0	1
1	1	1	1

Output Y is taken by random selection of 1. First we draw the blank map.

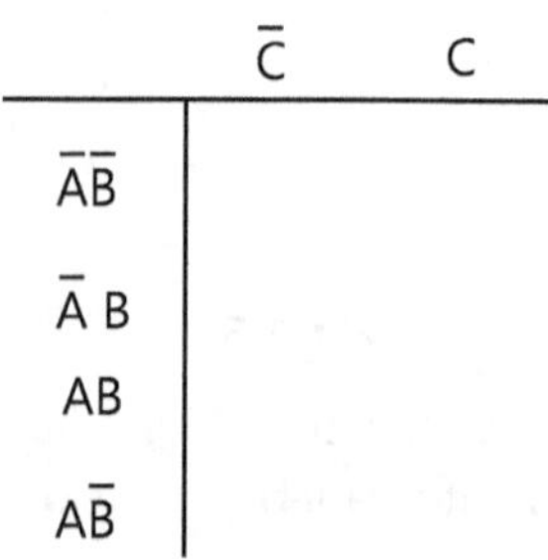

Fig. 1.60

The vertical column is labelled as $\overline{A}\overline{B}$, $\overline{A}B$, AB and A$\overline{B}$.

Output Y = 1 appears for 010, 110 and 111. The fundamental products for these input conditions are $\overline{A}$ B$\overline{C}$, AB$\overline{C}$, ABC (bar on variables where it is 0).

Now, enter 1s for these products in K-map.

	$\overline{C}$	C
$\overline{A}\overline{B}$		
$\overline{A}$ B	1	
AB	1	1
A$\overline{B}$		

Fig. 1.61

Enter 0s in the remaining spaces. Then final K-map becomes as in Fig. 1.62.

	$\overline{C}$	C
$\overline{A}\overline{B}$	0	0
$\overline{A}$ B	1	0
AB	1	1
A$\overline{B}$	0	0

Fig. 1.62

1.41 Pairs, Quads and Octets

(a) Pair : The K map shown in Fig. 1.63 contains a pair of 1s that are horizontally adjacent (next to each other).

(a)

	$\overline{C}\overline{D}$	$\overline{C}D$	CD	C$\overline{D}$
$\overline{A}\,\overline{B}$	0	0	0	0
$\overline{A}B$	1	1	0	0
AB	0	0	0	0
A$\overline{B}$	0	0	0	0

(b)

	$\overline{C}\overline{D}$	$\overline{C}D$	CD	C$\overline{D}$
$\overline{A}\,\overline{B}$	0	0	0	0
$\overline{A}B$	1	1	0	0
AB	0	0	0	0
A$\overline{B}$	0	0	0	0

(a) **(b)**

Fig. 1.63 : Horizontally adjacent ones

The first 1 represents the product of $\bar{A}\,B\bar{C}\bar{D}$ and second 1 the product of $\bar{A}\,B\bar{C}\,D$. As we move from the first 1 to second 1, only one variable goes from complemented to uncomplemented ($\bar{D}$ to D), the other variables do not change the form ($\bar{A}\,B\bar{C}$ remains unchanged). In such case, we can eliminate the variable that changes the form.

The sum of product (SOP) in which each individual term called as minterm represents

$$Y = \bar{A}\,B\bar{C}\bar{D} + \bar{A}\,B\bar{C}\,D$$

$$= \bar{A}\,B\bar{C}\,(\bar{D} + D) \qquad \because \bar{D} + D = D + \bar{D}$$

$$= \bar{A}\,B\bar{C}\,(D + \bar{D})$$

$$= \bar{A}\,B\bar{C} \qquad \because D + \bar{D} = 1$$

$$\therefore \qquad Y = \bar{A}\,B\bar{C}$$

Adjacent 1s as shown in Fig. 1.63 (a) complements are dropped out. For easy identification we will encircle a pair of adjacent 1s as shown in Fig. 1.64 (b).

For the pair of horizontally or vertically adjacent 1s, we can eliminate the variable that appears in both complemented and uncomplemented form.

Examples of pairs for 3 variables

	$\bar{C}$	C
$\bar{A}\,\bar{B}$	1	0
$\bar{A}\,B$	1	1
$A\,B$	0	0
$A\,\bar{B}$	0	0

Fig. 1.63 (c)

B goes from complemented to uncomplemented form ($\bar{B}$ to B).

$$Y = \bar{A}\bar{B}\bar{C} + \bar{A}\,B\bar{C}$$

$$= \bar{A}\bar{C}$$

If more than one pair exist on a Karnaugh map, we can OR the simplified products to get the Boolean equation.

For example,

	$\overline{C}\overline{D}$	$\overline{C}D$	CD	$C\overline{D}$
$\overline{A}\overline{B}$	1	1	0	0
$\overline{A}B$	0	0	0	0
AB	0	0	0	1
$A\overline{B}$	0	0	0	1

Fig. 1.63 (d)

For upper horizontal pair,

$$Y = \overline{A}\overline{B}\overline{C}\overline{D} + \overline{A}\overline{B}\overline{C}D$$

$$\therefore \qquad Y = \overline{A}\overline{B}\overline{C}$$

and for the lower vertical pair.

$$Y = AB C\overline{D} + A\overline{B}\,C\overline{D}$$

$$\therefore \qquad Y = AC\overline{D}$$

The corresponding Boolean equation for this map is

$$Y = \overline{A}\overline{B}\overline{C} + AC\overline{D}$$

(b) Quad : A quad is a group of four 1s that are horizontally or vertically adjacent. The 1s may be end to end or in the form of square.

This group can be formed by combining top row, bottom row, left column, right column, just like in pairing. Infact, quad eliminates two variables and their complements.

Consider K-map for three variables.

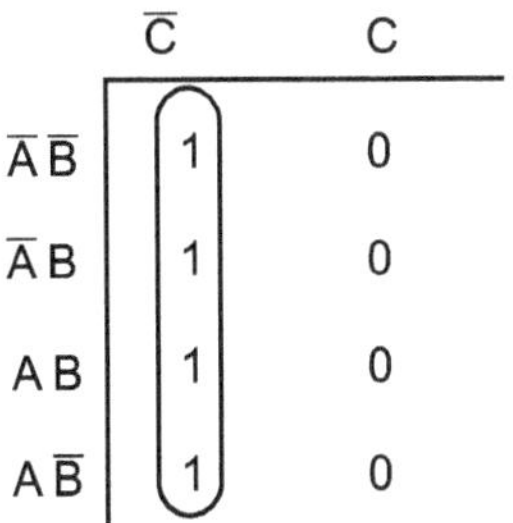

	$\overline{C}$	C
$\overline{A}\overline{B}$	1	0
$\overline{A}B$	1	0
AB	1	0
$A\overline{B}$	1	0

Fig. 1.64

Looking in K-map from Fig. 1.64, we find except for $\bar{C}$, other variables AB are changed from complement to uncomplement form and/or vice versa. Therefore, output Y becomes from Boolean algebra,

$$Y = \bar{A}\bar{B}\bar{C} + \bar{A}\,B\bar{C} + AB\bar{C} + A\bar{B}\bar{C}$$

$$Y = \bar{A}\bar{C}\,(\bar{B} + B) + A\bar{C}\,(B + \bar{B})$$

$$Y = \bar{A}\bar{C}\,(B + \bar{B}) + A\bar{C}\,(B + \bar{B}) \qquad\qquad (\because B + \bar{B} = 1)$$

$$Y = \bar{A}\bar{C} + A\bar{C}$$

$$Y = \bar{C}\,(\bar{A} + A)$$

$$Y = \bar{C} \qquad\qquad (\because \bar{A} + A = 1)$$

The other combinations of quads are

(a)

	$\bar{C}\bar{D}$	$\bar{C}D$	CD	$C\bar{D}$
$\bar{A}\bar{B}$	0	0	0	0
$\bar{A}B$	1	1	1	1
AB	0	0	0	0
$A\bar{B}$	0	0	0	0

Fig. 1.65

$$Y = \bar{A}\,B\bar{C}\bar{D} + \bar{A}\,B\bar{C}\,D + \bar{A}\,BCD + \bar{A}\,BC\bar{D}$$

$$= \bar{A}\,B\bar{C}\,(\bar{D} + D) + \bar{A}\,BC\,(D + \bar{D})$$

$$= \bar{A}\,B\bar{C} + \bar{A}\,BC \qquad\qquad (\because D + \bar{D} = 1)$$

$$= \bar{A}\,B\,(\bar{C} + C) \qquad\qquad (\because \bar{C} + C = 1)$$

$$= \bar{A}\,B$$

(b)

	$\overline{C}\,\overline{D}$	$\overline{C}\,D$	$C\,D$	$C\,\overline{D}$
$\overline{A}\,\overline{B}$	0	0	0	0
$\overline{A}\,B$	0	0	1	1
$A\,B$	1	1	0	0
$A\,\overline{B}$	1	1	0	0

Fig. 1.66

$$Y = AB\overline{C}\,\overline{D} + AB\overline{C}\,D + A\overline{B}\,\overline{C}\,\overline{D} + A\overline{B}\,\overline{C}\,D$$

$$= AB\overline{C}\,(\overline{D} + D) + A\overline{B}\,\overline{C}\,(\overline{D} + D)$$

$$= AB\overline{C} + A\overline{B}\,\overline{C}$$

$$= A\overline{C}\,(B + \overline{B}) \qquad\qquad (\because B + \overline{B} = 1)$$

$$= A\overline{C}$$

(c)

	$\overline{C}\,\overline{D}$	$\overline{C}\,D$	$C\,D$	$C\,\overline{D}$
$\overline{A}\,\overline{B}$	0	0	0	0
$\overline{A}\,B$	1	0	0	1
$A\,B$	1	0	0	1
$A\,\overline{B}$	0	0	0	0

Fig. 1.67

$$Y = \overline{A}\,B\overline{C}\,\overline{D} + \overline{A}\,BC\overline{D} + AB\overline{C}\,\overline{D} + ABC\overline{D}$$

$$Y = \overline{A}\,B\overline{D}\,(\overline{C} + C) + AB\overline{D}\,(\overline{C} + C)$$

$$Y = \overline{A}\,B\overline{D} + AB\overline{D} \qquad\qquad (\because \overline{C} + C = 1)$$

$$Y = B\overline{D}\,(\overline{A} + A)$$

$$Y = B\overline{D}$$

(c) Octet : This is a group of eight adjacent 1s. An octet like this eliminates three variables and their complements. Octet can be considered as a pair of quads.

(i)   Consider the K-map shown in Fig. 1.68.

	$\overline{C}\,\overline{D}$	$\overline{C}\,D$	$C\,D$	$C\,\overline{D}$
$\overline{A}\,\overline{B}$	0	0	0	0
$\overline{A}\,B$	1	1	1	1
$A\,B$	1	1	1	1
$A\,\overline{B}$	0	0	0	0

Fig. 1.68

The Boolean expression will be

$$Y = \overline{A}\,B\overline{C}\overline{D} + \overline{A}\,B\overline{C}\,D + \overline{A}\,BCD + \overline{A}\,BC\overline{D}$$

$$+ AB\overline{C}\overline{D} + AB\overline{C}\,D + ABCD + ABC\overline{D}$$

$\therefore$
$$Y = \overline{A}B\overline{C}\,(\overline{D} + D) + \overline{A}BC\,(D + \overline{D}) + AB\overline{C}\,(\overline{D} + D)$$

$$+ ABC\,(D + \overline{D})$$

$$Y = \overline{A}B\overline{C} + \overline{A}BC + AB\overline{C} + ABC$$

$$= \overline{A}B\,(\overline{C} + C) + AB\,(\overline{C} + C)$$

$$= \overline{A}B + AB = B\,(\overline{A} + A) = B$$

(ii)   Consider the K-map shown in Fig. 1.69.

	$\overline{C}\,\overline{D}$	$\overline{C}\,D$	$C\,D$	$C\,\overline{D}$
$\overline{A}\,\overline{B}$	0	0	0	0
$\overline{A}\,B$	0	0	0	0
$A\,B$	1	1	1	1
$A\,\overline{B}$	1	1	1	1

Fig. 1.69

$$Y = AB\overline{C}\overline{D} + AB\overline{C}\,D + ABCD + ABC\overline{D}$$

$$+ A\overline{B}\overline{C}\overline{D} + A\overline{B}\overline{C}\,D + A\overline{B}\,CD + A\overline{B}\,C\overline{D}$$

First eliminate D,

$$Y = AB\overline{C}\,(\overline{D} + D) + ABC\,(D + \overline{D}) + A\overline{B}\overline{C}\,(\overline{D} + D)$$

$$+ A\overline{B}C\,(D + \overline{D})$$

$$= AB\overline{C} + ABC + A\overline{B}\overline{C} + A\overline{B}\,C$$

Now eliminate C,

$$Y = AB(\bar{C} + C) + A\bar{B}(\bar{C} + C)$$

$$= AB + A\bar{B}$$

Now eliminate B,

$$Y = A(B + \bar{B})$$

$$Y = A$$

In this way, three variables B, C, D and their complements dropout from the corresponding product.

1.42 Introduction to ALU

- ALU is an abbreviation of arithmetic logic unit, the part of a computer that performs all arithmetic computations, such as addition and multiplication, and all comparison operations. The ALU is one component of the CPU (central processing unit) of the computer. Mathematician John von Neumann proposed the ALU concept in 1945, when he wrote a report on the foundations for a new computer called the EDVAC.

- Arithmetic logic unit is a digital circuit which performs both arithmetic and logical operations. The basic arithmetic operation performed by ALU is addition, however using this further, subtraction, multiplication and division is also obtained.

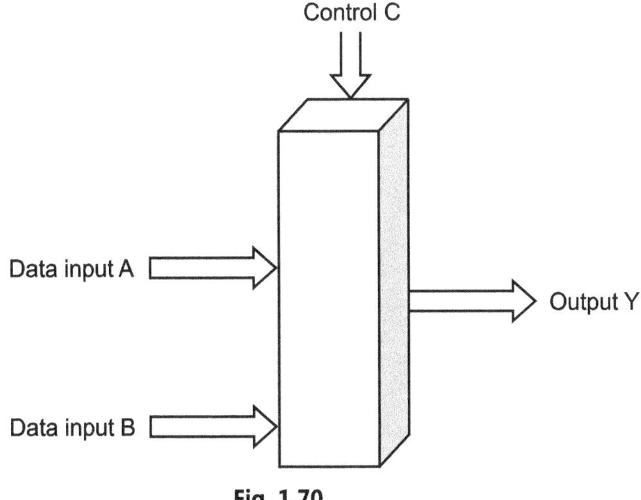

Fig. 1.70

- The logic operations are usually Not (Inversion), AND, OR and EX-OR. The ALU can be represented as shown in Fig. 1.70. ALU has two data inputs A and B and an output Y as shown in Fig. 1.70. The digital levels on control input C will determine which operation is to be performed.

- The width of inputs A, B output Y and control input C depends upon data width to be operated and hence is referred to as bus.

- ALU can be obtained using discrete logic gates. However to make designer's job easy, ALU are also available in the form of ICs. IC 74181 is an ALU, the first complete ALU on a single chip. It was used as the arithmetic/logic core in the CPUs of many historically significant minicomputers and other devices.

- The IC 74181 represents an evolutionary step between the CPUs of the 1960s, which were constructed using discrete logic gates, and today's single-chip CPUs or microprocessors.

- The 74181 is a 7400 series medium-scale integration (MSI) TTL integrated circuit, containing the equivalent of 75 logic gates and most commonly packaged as a 24-pin DIP. IC 74181 is shown in Fig. 1.71.

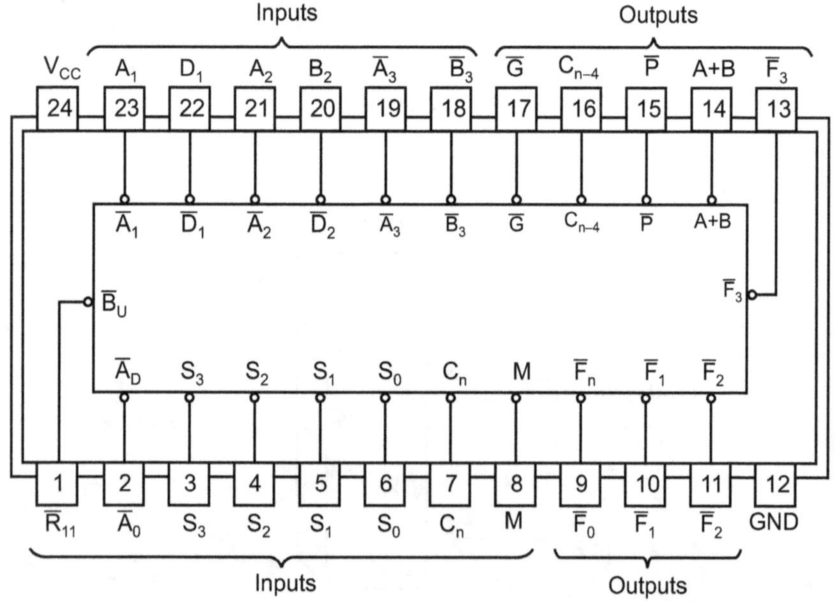

Fig. 1.71 : IC 74181

- The 4-bit wide ALU can perform all the traditional add / subtract / decrement operations with or without carry, as well as AND / NAND, OR / NOR, XOR, and shift. Many variations of these basic functions are available, for a total of 16 arithmetic and 16 logical operations on two four-bit words.

- Multiply and divide functions are not provided but can be performed in multiple steps using the shift and add or subtract functions.

Solved Examples

Example 1.5 : *Draw a logic circuit and obtain truth table for the following expression* $Y = A + \overline{(B \cdot C)} + \overline{A \cdot B} + C.$

Solution : Consider the Fig. 1.72.

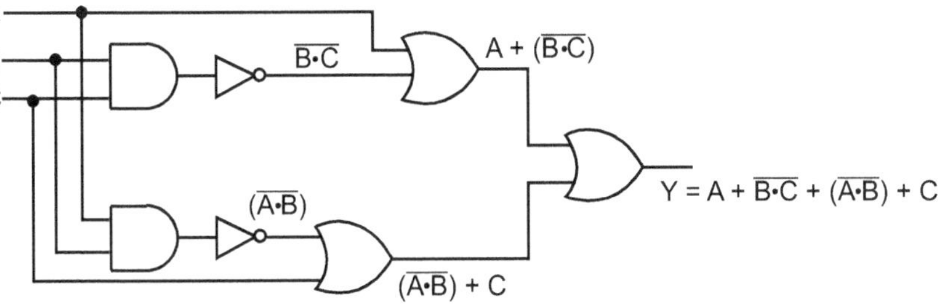

Fig. 1.72

Table 1.16 : Truth table

A	B	C	Y
0	0	0	1
0	0	1	1
0	1	0	1
0	1	1	1
1	0	0	1
1	0	1	1
1	1	0	1
1	1	1	1

Using Boolean expression, we can reduce this circuit as follows.

Consider $Y = A + \overline{B \cdot C} + \overline{A \cdot B} + C$

Applying De Morgan's theorem,

$$Y = A + \bar{B} + \bar{C} + \bar{A} + \bar{B} + C \qquad (\because \overline{B \cdot C} = \bar{B} + \bar{C})$$

$$\therefore \quad Y = (A + \bar{A}) + \bar{B} + \bar{B} + C + \bar{C} \qquad (\overline{A \cdot B} = \bar{A} + \bar{B})$$

$$= 1 + \bar{B} + 1$$

$$Y = 1 + \bar{B} \qquad (\because A + \bar{A} = 1)$$

$$Y = 1 \qquad (\because C + \bar{C} = 1)$$

Thus, output Y is HIGH irrespective of any input H or L as it can be seen from truth table also.

Example 1.6 : *Simplify the equation and then draw logic diagram.*

$$Y = \overline{A}\,\overline{B}\,\overline{C} + \bar{A}\,B\bar{C} + A\overline{B}\,\overline{C} + AB\bar{C}$$

Solution : Consider $Y = \overline{A}\,\overline{B}\,\overline{C} + \bar{A}B\bar{C} + A\overline{B}\,\overline{C} + AB\bar{C}$

$$\therefore \quad Y = \overline{A}\,\overline{C}\,(\bar{B} + B) + A\bar{C}\,(\bar{B} + B)$$

$$= \overline{A}\,\overline{C}\,(1) + A\bar{C}\,(1) \qquad (\because \bar{B} + B = 1)$$

$$= \overline{A}\,\overline{C} + A\bar{C}$$

$$= \bar{C}\,(\bar{A} + A) \qquad (\because \bar{A} + A = 1)$$

$$= \bar{C}$$

The logic circuit to solve the above equation is shown in Fig. 1.73.

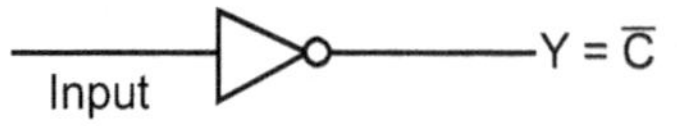

Fig. 1.73

We can verify the result by considering the input conditions : A = 0, B = 0 and C = 1.

The expected result is $Y = \bar{C}$ i.e. Y = 0

By applying the input, we get,

$$Y = \overline{A}\,\overline{B}\,\overline{C} + \bar{A}\,B\bar{C} + A\overline{B}\,\overline{C} + AB\bar{C}$$

$$= \overline{0}\,\overline{0}\,\overline{1} + \overline{0}\,0\overline{1} + 0\,\overline{0}\,\overline{1} + 0\,0\,\overline{1}$$

$$= 0 + 0 + 0 + 0$$

$$= 0$$

$$\text{i.e. } Y = \bar{C}$$

Example 1.7 : *Simplify the following equation and then draw logic diagram and truth table.*

$$Y = A\bar{B}C + A\bar{B}\bar{C} + B$$

Solution : Consider

$$Y = A\bar{B}C + A\bar{B}\bar{C} + B$$

$\therefore \qquad Y = A\bar{B}(C + \bar{C}) + B$

$\qquad\qquad = A\bar{B}(1) + B \qquad\qquad (\because C + \bar{C} = 1)$

$\qquad\qquad = A\bar{B} + B$

$\qquad\qquad = A + B \qquad\qquad (\because A\bar{B} + B = A + B)$

The logic circuit is,

$$Y = A + B$$

Fig. 1.74

Table 1.17 : Truth table

Inputs		Output
A	**B**	**Y = A + B**
0	0	0
0	1	1
1	0	1
1	1	1

Example 1.8 : *Simplify the following Boolean equation and then draw logic diagram and truth table :*

$$Y = AB\bar{C} + ABC + BC$$

Solution : $\qquad Y = AB\bar{C} + ABC + BC$

$\qquad\qquad Y = AB(\bar{C} + C) + BC \qquad\qquad (\because C + \bar{C} = 1)$

$\qquad\qquad Y = AB(1) + BC$

$\qquad\qquad Y = AB + BC$

$\qquad\qquad Y = B(A + C)$

The logic circuit is,

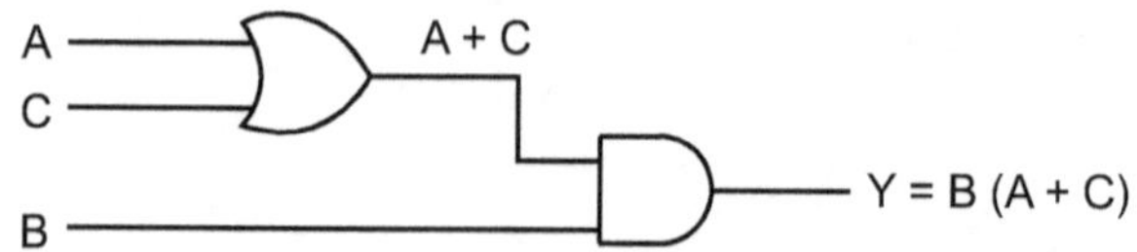

Fig. 1.75

Table 1.18 : Truth table

Inputs			Output
A	**B**	**C**	**Y = B (A + C)**
0	0	0	0
0	0	1	0
0	1	0	0
0	1	1	1
1	0	0	0
1	0	1	0
1	1	0	1
1	1	1	1

Example 1.9 : *Minimise the equation* $y = \bar{A}\bar{B}C + \bar{A}BC + A\bar{B}C + ABC$ *using Boolean Algebra or K-maps.*

Solution : (i) Using Boolean algebra : Rearranging the above

$$y = \bar{A}\bar{B}C + A\bar{B}C + \bar{A}BC + ABC$$

$$= (\bar{A} + A)\,\bar{B}C + (\bar{A} + A)\,BC \qquad (\because A + \bar{A} = 1)$$

$$= \bar{B}C + BC = C(\bar{B} + B)$$

$$y = C$$

$$\therefore \qquad y = \bar{A}\bar{B}C + \bar{A}BC + A\bar{B}C + ABC$$

(ii) Using K-map : Since equation contains three variables, the K-map will have $2^3 = 8$ cells.

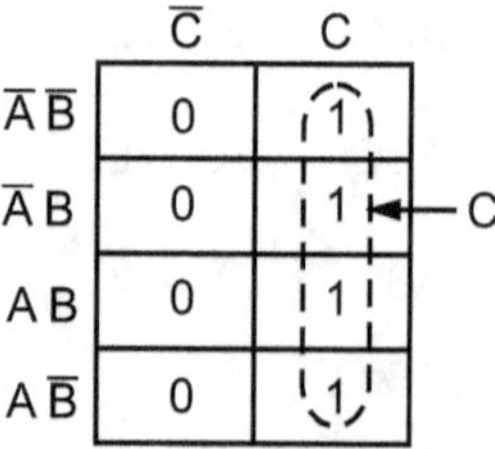

Fig. 1.76

$$y = C$$

Example 1.10 : *Draw the logic diagram for the expression*

$$y = \overline{(\overline{A} \cdot B + \overline{B} \cdot \overline{C})}.$$

Solution :

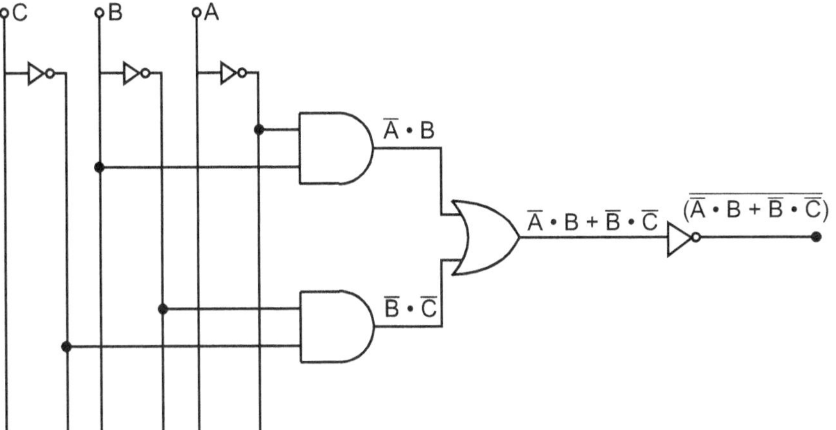

Fig. 1.77

Example 1.11 : *Simplify the following SOP expression using K-map.*

$$Y = ABC + \overline{A}\,\overline{B}\,\overline{C} + AB\overline{C} + \overline{A}BC$$

Solution : The K-map for the given equation is

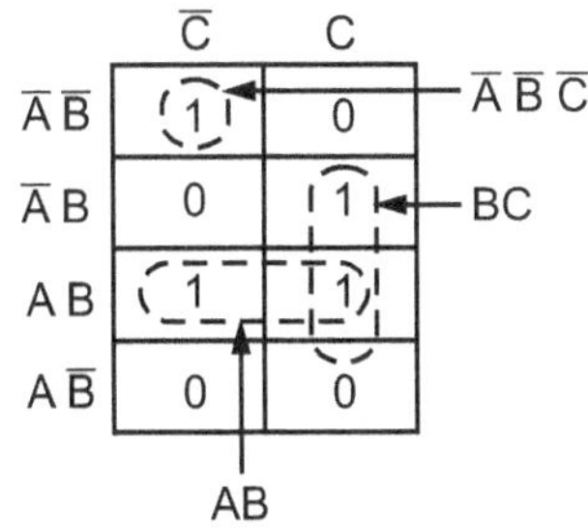

Fig. 1.78

$$\therefore \qquad y = AB + BC + \overline{A}\,\overline{B}\,\overline{C}$$

Example 1.12 : *Simplify the logic expression.*

$y = \overline{A}B + \overline{A}\,\overline{B}\,\overline{C} + \overline{A}BCD + \overline{A}BC\overline{D}$ *using the laws of Boolean algebra. Draw simplified logic diagram.*

Solution : Consider expression

$$y = \bar{A}B + \bar{A}B\bar{C} + \bar{A}BCD + \bar{A}BC\bar{D}$$

$$= \bar{A}B (1 + \bar{C}) + \bar{A}BC (D + \bar{D}) \quad (\because 1 + \bar{C} = 1, D + \bar{D} = 1)$$

$$= \bar{A}B + \bar{A}BC$$

$$= \bar{A}B (1 + C)$$

$$= \bar{A}B \qquad\qquad (\because 1 + C = 1)$$

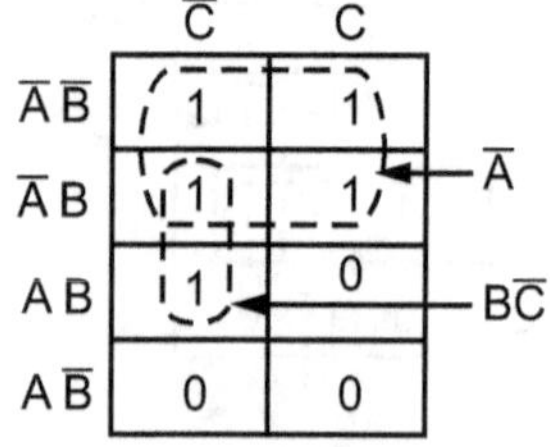

Fig. 1.79 : Logic diagram

Example 1.13 : *Minimize the following logical expression using K-maps.*

$$y = \bar{A}\bar{B}C + \bar{A}\bar{B}\bar{C} + AB\bar{C} + \bar{A}B\bar{C} + \bar{A}BC$$

Solution : Consider expression $y = \bar{A}\bar{B}C + \bar{A}\bar{B}\bar{C} + AB\bar{C} + \bar{A}BC$

Since expression has three variables so K-map will have $2^3 = 8$ cells.

	$\bar{C}$	C	
$\bar{A}\,\bar{B}$	1	1	
$\bar{A}\,B$	1	1	— $\bar{A}$
$A\,B$	1	0	— $B\bar{C}$
$A\,\bar{B}$	0	0	

Fig. 1.80

$$\therefore \qquad y = \bar{A} + B\bar{C}$$

Example 1.14 : *Convert the following SOP expression into standard*
SOP form : Y =   AB + AC + B\bar{C}

Solution : Given expression is

$$Y = AB + AC + B\bar{C}$$

Step 1 : Find the missing literal in each term

$$Y = AB + AC + B\bar{C}$$

↑	↑	↑
missing	missing	missing
literal is	literal is	literal is
C	B	A

Step 2 : And each term with its (missing literal + its complement)

$$Y = AB (C + \bar{C}) + AC (B + \bar{B}) + B\bar{C} (A + \bar{A})$$

Since $C + \bar{C} = 1$, the value of expression does not change as

$$Y = ABC + AB\bar{C} + ABC + A\bar{B}C + AB\bar{C} + \bar{A}B\bar{C}$$

$$= ABC + ABC + AB\bar{C} + AB\bar{C} + A\bar{B}C + \bar{A}B\bar{C} \qquad (\because A + A = A)$$

$$= ABC + AB\bar{C} + A\bar{B}C + \bar{A}B\bar{C}$$

Since in the above expression each term contains all the literals it is in the standard SOP form.

Example 1.15 : *Simplify the following expression using K-map :*

$$Y = \bar{A}\bar{B}\bar{C} + \bar{A}\bar{B}C + \bar{A}BC + ABC$$

Solution   :   Given   expression contains three literals. So the K-map will have $2^3 = 8$ cells.

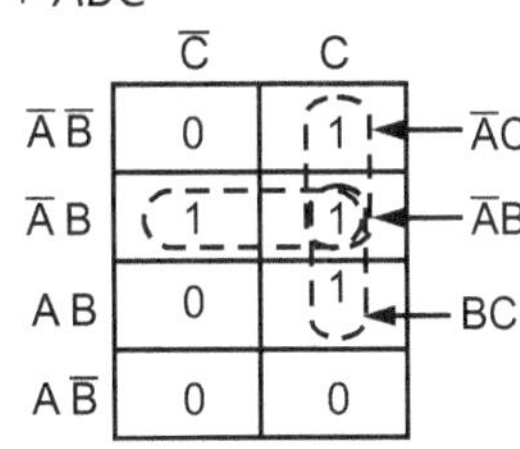

Fig. 1.81

$$y = \bar{A}C + \bar{A}B + BC$$

Example 1.16 : *Simply the following using Boolean algebra.*

$$\bar{A}\bar{B}C + \overline{(A + B + \bar{C})} + \overline{\bar{A}\bar{B}}CD.$$

Solution : $Y = \bar{A}\bar{B}C + \bar{A} \cdot \bar{B} \cdot \bar{\bar{C}} + \overline{\bar{A}\bar{B}}CD$ Using $\overline{A + BC} = \bar{A} \cdot \bar{B} \cdot \bar{C}$

$$= \bar{A}\bar{B}C + \bar{A} \cdot \bar{B} \cdot C + \overline{\bar{A}\bar{B}}CD \qquad\qquad \text{Using } \bar{\bar{A}} = A$$

$$= \bar{A}\bar{B}C + \overline{\bar{A}\bar{B}}CD \qquad\qquad (\because A + A = A)$$

$$Y = \bar{A}\bar{B} (C + \bar{C}D)$$

Think Over It

1. In day to day life we come across more analog thing rather than digital.
2. For entering data into circuit we have to use binary data.
3. In certain situation we have to use –ve logic system.
4. Three input AND gate can be used as two input AND gate.
5. Universal gates are sufficient to build any logic circuit.
6. Boolean algebra helps to simplify logic equations.
7. Karnaugh map is more systematic way to simplify logic expressions.
8. Quad in K-map eliminates two variables.

Points to Remember

1. Boolean algebra is the mathematics of digital system.
2. The Boolean axioms are the laws of Boolean algebra for addition, multiplication and inversion.
3. The Boolean operators include Dot sign (·) which indicates logical product of two terms, Plus sign (+) which represents logical sum of terms and Overbar (¯) indicates that the terms which are overbar, are to be complemented.
4. Boolean addition is the same as the OR.
5. Boolean multiplication is the same as the AND.
6. Logic equations are simplified using laws of Boolean algebra.
7. We can reduce number of gates required for designing a digital circuit, reduce a particular expression to its simplest form, ultimately which reduces size and price (cost) of circuit by applying basic laws, rules and theorems of Boolean algebra.
8. Boolean expression can be converted into either of two standard forms; the sum of product (SOP) form or the product of sum (POS) form.
9. When two or more product terms are summed by Boolean addition, the resulting expression is known as sum of product (SOP) form.
10. When two or more sum terms are multiplied, the resulting expression is known as product of sum (POS) form.
11. De Morgan's first theorem : The complement of product is equal to the sum of the complements. It can be expressed as,

$$\overline{AB} = \overline{A} + \overline{B}$$

12. De Morgan's Second Theorem : The complement of a sum is equal to the product of the complements. It can be expressed as,

$$\overline{A + B} = \overline{A} \cdot \overline{B}$$

13. The NAND and NOR gates have universal property and can be used for performing AND, OR and NOT functions, so an AND/OR/NOT logic circuit can be converted to NAND/NOR logic.

14. The Karnaugh map or K-map provides a systematic method for simplifying a Boolean expression and can be used as visual display of fundamental products needed for a sum of product solution.

15. K-map provides easiest way to produce simplest Boolean expression with minimum chances of error.

16. Karnaugh maps are used to simplify digital logic circuits so that they can be implemented using a minimum number of physical logic gates.

17. Analog quantity is one having continuous value, whereas digital quantity has discrete set of values.

18. The basic building block used in digital circuits operate in one of the two states, known as ON and OFF or '1' and '0' which results in a very simple operation.

19. A number system is an ordered set of symbols known as digits and the total number of symbols used in a system defines the type of number system.

20. Some of the commonly used number systems in digital circuits are binary, octal and hexadecimal number system.

21. Decimal number system is ordered set of ten symbols 0, 1, 2, 3, 4, 5, 6, 7, 8 and 9 are used to specify the quantities.

22. The decimal number can be expressed as sum of powers of ten.

23. In computers, logic circuits, microprocessors etc. the commonly used number systems are binary, octal and hexadecimal number system.

24. The main advantage of using binary number system is that it minimizes the number of lines in a computer system. A binary digit is known as 'bit' in digital representation and the group of four bits is known as nibble, group of eight bits is known as *byte*.

25. The binary number system is a code that uses only two basic symbols i.e. 0 and 1.

26. In a binary number, the left most bit is known as most significant bit (MSB) and the right most bit is known as least significant bit (LSB). MSB has highest weight while LSB has lowest weight.

27. Any binary number can be converted into its equivalent decimal number using the weights assigned to each bit position of the order 1, 2, 4, 8, 16....from LSB to MSB.

28. A popular way to convert decimal to binary numbers is double-dabble method or divide by two method.

29. Hexadecimal number system has base of sixteen. It is extensively used in microprocessor work. Hexadecimal number represents group of four bit binary numbers. It uses 16 distinct symbols 0 to 9 and A to F. Thus, the symbols in hexadecimal number systems are 0, 1, 2, 3, 4, 5, 6, 7, 8, 9, A, B, C, D, E, F.

30. A subscript 16 or H is used which indicates a hexadecimal number. Hexadecimal number can be converted into binary number by converting each hexadecimal digit to 4-bit binary equivalent. Binary number can be easily converted to hexadecimal by making a group of 4-bit starting from binary point. Then group is converted to equivalent hexadecimal.

31. In the hexadecimal number system each digit position corresponds to power of 16.

32. The decimal number can be converted to hexadecimal number by successively dividing by 16. Then the remainder is converted to the equivalent hexadecimal. The remainder taken in reverse order, written from the bottom to the top, gives hexadecimal number.

33. The BCD code expresses each digit in a decimal number by its 4-bit binary equivalent.

34. A signed binary number consists of both sign and magnitude information.

35. There are three forms in which signed integer numbers can be represented in binary, they are sign magnitude, 1's complement and 2's complement form.

36. In a binary number, if each 1 is replaced by 0 and each 0 by 1, the resulting number is known as one's complement of the first number. The two's complement of binary number is obtained by adding 1 to 1's complement.

37. Non-weighted binary codes do not follow the positional weighting principle, i.e., digit position within the number has not assigned fixed value. Examples of non-weighted code are gray code, excess-3 code and alphanumeric code etc.

38. The ASCII (American Standard Code for Information Interchange) code is extensively used for data communication and in computers.

39. The excess-3 is 4-bit code sometimes used with BCD number. It is derived by adding 3 to each decimal digit and then converting the result to the 4-bit binary. Since no definite weight can be assigned to the four digit position, excess-3 is an unweighted code.

40. The Gray code is a binary numeral system where two successive values differ in only one bit.

41. A logic gate is a device implementing a Boolean function, that is, it performs a logical operation on one or more logical inputs, and produces a single logical output.

42. In a digital system, there are two discrete levels HIGH (1) and LOW (0). If the higher of the two voltages represents a 1 and lower voltage represents a 0, the system is called *positive logic system*. On the other hand, if lower voltage represents a 1 and higher voltage represents a 0 we have negative logic system.

43. The basic logic gates are AND, OR, NOT etc. These basic gates can be combined to perform other important logic operations like NAND, NOR and EX-NOR gates. So these are called as *derived gates*.

44. Any Boolean (or logic) expression can be realized by using the AND, OR and NOT gates. NAND, NOR operations can be derived from it hence NAND and NOR gates are known as *universal gates*.

45. The NOT gate is also called inverter as it inverts the input signal.

46. Information is normally handled by a digital system in a group of bits which is called as *word*. A parity bit is attached to group of information bits in order to make the total number of 1s even or odd. Even parity means an n-bit input has an even number of 1s whereas odd parity means an n-bit input has an odd number of 1s.

47. Exclusive-OR (XOR) gate has high output only when an odd number of input is high i.e. if single input is 1, output is 1. Therefore, XOR gates are ideal for checking the parity of binary number.

Exercise

[A] Multiple Choice Questions - I :

1. According to the AND laws A . A = ____
   (a) 0
   (b) 1
   (c) A
   (d) none of these

2. According to the OR laws A + $\bar{A}$ = ________
   (a) 0
   (b) 1
   (c) A
   (d) A

3. De Morgan's theorem says that $\overline{AB}$ = _____

   (a) A + B
   (b) $\bar{A} + \bar{B}$

   (c) $\overline{A + B}$
   (d) AB

4. A + AB = ____
   (a) B
   (b) AB
   (c) A
   (d) none of these

5. In Karnaugh-map, a pair eliminates _____ variables.
   (a) 1
   (b) 2
   (c) 3
   (d) 4

6. In Karnaugh-map, a quad eliminates _____ variables.
   (a) 1
   (b) 2
   (c) 3
   (d) 4

7. In Karnaugh-map, an octet eliminates _____ variables.
   (a) 1
   (b) 2
   (c) 3
   (d) 4

8. Each individual term in standard SOP form is called as _______.
   (a) minterm
   (b) maxterm
   (c) constant term
   (d) none of these

9. Each individual term in standard POS form is called as _______.
   (a) minterm
   (b) maxterm
   (c) constant term
   (d) none of these

Answers

1. (c)	2. (b)	3. (b)	4. (c)	5. (a)	6. (b)	7. (c)	8. (a)	9. (b)

Multiple Choice Questions - II :

1. Binary numbers are to the base _______
   (a) 8
   (b) 10
   (c) 2
   (d) 16

2. The hexadecimal numbers are to the base of _______
   (a) 10   (b) 16
   (c) 2   (d) 8

3. Conversion of $(365.26)_8$ into its equivalent binary number is _____
   (a) $(1010011.01100111)_2$ (b) $(1011011.010101)_2$
   (c) $(011110101.010110)_2$ (d) None of these

4. Conversion of $(68.4C)_{16}$ into its equivalent binary number is _____
   (a) $(0110011.01100111)_2$ (b) $(01101000.01001100)_2$
   (c) $(011110101.010110)_2$ (d) None of these

5. Range of sign-magnitude numbers are _____
   (a) -127 to $+127$ (b) -128 to $+128$
   c) -256 to $+256$ (d) -255 to $+255$

6. Conversion of $(99)_{10}$ into its equivalent BCD number is _____
   (a) $(10011000)_{BCD}$ (b) $(10011001)_{BCD}$
   (c) $(00011001)_{BCD}$ (d) None of these

7. The largest number is _______
   (a) $(1001)_{BCD}$ (b) $(1111)_{BCD}$
   (c) $(1111111)_{BCD}$ (d) None of these

8. The decimal numbers which can be expressed in BCD are from _______
   (a) 0 to 15   (b) 0 to 9
   (c) 0 to 255   (d) None of these

9. The example of weighted codes is _____
   (a) ASCII   (b) Gray
   (c) Hexadecimal   (d) None of these

10. The example of non-weighted codes is _____
    (a) Excess-3   (b) Binary
    (c) Decimal   (d)   Hexadecimal

11. Conversion of $(1110)_{Gray}$ into its equivalent binary number is _____
    (a) $(1001)_{BCD}$ (b) $(0101)_{BCD}$
    (c) $(1011)_{BCD}$ (d) None of these

12. The long form of ASCII is _____
    (a) American Standard Code for Information Interchange
    (b) American Standard Code for Internet Interchange
    (c) American Standard Code for Interchange Information
    (d) None of these

13. Conversion of $(01100011)_{BCD}$ into its equivalent decimal number is _____
    (a) $(63)_{10}$ (b) $(36)_{10}$
    (c) $(163)_{10}$ (d) None of these

14. Conversion of $(01010011)_{BCD}$ into its equivalent Excess-3 number is ____

   (a) $(10000110)_{Ex-3}$ (b) $(00100001)_{Ex-3}$
   (c) $(101011100)_{Ex-3}$ (d) None of these

15. ____ type of codes are not used for performing arithmetic operations.
   (a) Decimal   (b) Binary
   (c) Octal   (d) Gray

16. The diode is used to ____ two devices.
   (a) Couple   (b) Isolate
   (c) Link   (d) None of these

17. In negative logic system logic 0 is expressed as ____
   (a) 0 V   (b) 5 V
   (c) 4.5 V   (d) 2.5 V

18. The Boolean expression for Ex-OR gate is ____

   (a) $\overline{A}B + BA$ (b) $\overline{A}B + A\overline{B}$
   (c) $A + B$ (d) None of these

19. ____ is/are called as universal gates.
   (a) NAND   (b) NOR
   (c) AND   (d) both (a) and (b)

20. In positive logic system logic 1 is expressed as ____
   (a) 0 V   (b) 4.5 V   (c) 5 V   (d) 3.5 V

21. ____ parity means an n-bit input has an even number of 1s.
   (a) Odd   (b) Even
   (c) Mixed   (d) None of these

22. The Boolean expression for OR gate is ____
   (a) $A + B$ (b) $A \cdot B$
   (c) $AB + BA$ (d) None of these

Answers

1. (c)	2. (b)	3. (c)	4. (b)	5. (a)	6. (b)	7. (a)	8. (b)
9. (c)	10. (a)	11. (c)	12. (a)	13. (a)	14. (a)	15. (d)	16. (b)
17. (b)	18. (b)	19. (d)	20. (c)	21. (b)	22. (a)		

[B] True Or False :

1. Digital quantity will base infinite values within the set of values.
2. Digital representation of quantity is more convenient than analog representation.
3. Decimal numbers are not suitable in electronics circuits.

4. BCD numbers does not obey the rules of binary addition.
5. Some systems of digital electronics uses −ve logic.
6. Gray code is weighted code of system.
7. Only NAND gates are sufficient to build any logic circuit.
8. Quad in K-map eliminates one variable.
9. Transistor can be used as NOT gate.
10. Four input NAND gate can be used as two input NAND gate.

Answers						
1. False	2. True	3. True	4. True	5. True	6. False	7. True
8. False	9. True	10. True				

[C] Short Answer Questions:

1. What is excess-3 code ?
2. What is a gray code ? What are its main characteristics ?
3. What is alphanumeric code ?
4. What is difference between binary and BCD number system ?
5. What is positive and negative logic level ? Explain with ideal pulse.
6. What is logic gate ? What are different types of basic gate ?

[D] Long Answer Questions:

1. State various laws of Boolean algebra.
2. State and verify De Morgan's 1st and 2nd theorems.
3. Explain procedure for converting a logic circuit into NAND logic.
4. Explain how logic circuit can be converted to NOR logic circuit.
5. What is K-map ? Where is it used ? What are its advantages ? Explain 3 variable K-map with suitable example.
6. Draw logic symbols and truth tables for NAND, NOR, EX-OR and NOT gate.
7. Simplify following equations using laws of Boolean algebra :

   (i) $Y = AB + BC + \bar{B}A + \bar{A}B$,   (ii) $Y = \bar{A} + AB + A\bar{B}$

   (iii) $Y = AB + \bar{A}B + ABC$

8. Write a note on logic families.
9. Why NAND gate is called as universal building block ? Explain it with suitable example.
10. Use only NOR gate to build NAND, OR and EX-OR gates.
11. Design all basic gates using NOR gate.

12. Convert the following expressions into their standard SOP and POS

    forms : (a) $Y = AB + AC + BC$ (b) $Y = (A + B) (\bar{B} + C)$
    (c) $Y = A + B + C + ABC$

13. What is the radix used in the case of decimal, binary, octal and hexadecimal ?

14. Write a note on BCD code.

15. Distinguish between binary code from BCD code.

16. Write notes on : ASCII code

17. Explain gray code system with suitable example.

18. Convert the following binary numbers into decimal and hexadecimal.
    (a) (i)   1101, (ii) 110010, (iii) 10101010, (iv) 11111111,
         (v) 11011011, (vi) 111000
    (b) (i) 1101.11, (ii) 1010.001,   (iii) 1001.111.

19. Convert the following decimal numbers into binary and hexadecimal.
    (a) (i) 95, (ii) 123, (iii) 221, (iv) 252, (v) 529, (vi) 445.
    (b) (i) 33.87,   (ii) 89.85, (iii) 107.23.

20. Convert the following hexadecimal numbers into decimal, binary.
    (a) (i)   1E, (ii) ABC, (iii) F2, (iv) A99, (v) 23C, (vi) C92.
    (b) (i) AE.35, (ii) 98.B2, (iii) C1.77.

21. Convert the following numbers in BCD code.
    (i)   72, (ii) 365, (iii) 593

22. Convert the following numbers to gray code.
    (i)   01001, (ii) 101101, (iii) 10110110.

23. Solve the following
    (i) $(77)_{10} = (?)_2$, (ii) $(99)_{10} = (?)_{16}$, (iii) $(1011101) = (?)_{16}$
    (iv) $(FA)_{16} = (?)_{10}$

24. Describe following gates with symbol and truth table :
    (a) OR gate, (b) AND gate, (c) NOT gate, (d) NOR gate,
    (e) NAND gate.

25. Why NAND and NOR gates are called as universal gate ? Explain.

26. How basic gates can be constructed by using NAND and NOR gate ?

27. Explain EX-OR gates with symbol and truth table.

28. Explain how EX-OR gate can work as parity generator.

29. Explain the circuit diagram for basic NOT gate.

30. Draw and explain the NAND gate circuit with transistor. Write a truth table for it.

31. Explain how OR gate circuit can be constructed using transistor. Write truth table for it.

32. Draw and explain the AND gate circuit with transistor. Write truth table for it.

Chapter **2**...

Combinational Logic Circuits

Blue, green, and red LEDs in 5 mm diffused case
Working principle
    Electroluminescence
Invented
    H. J. Round (1907)
    Oleg Losev (1927)
    James R. Biard (1961)
    Nick Holonyak (1962)
First production
    October 1962
Pin configuration
    Anode and cathode

	Captain Henry Joseph Round

2.1 Introduction

- Using logic gates, we can realize logical expressions which can be simplified using laws of Boolean algebra (we have studied logic gates in chapter 1). Digital circuits built using logic gates are classified into two categories:

  (a) Combinational circuits. (b) Sequential circuits.

- There are certain circuits whose output at any instant of time depends on input present at that instant of time and does not depends on any other factors, these circuits are known as *combinational circuits.*

- In such circuits, output is function of input, the function depends upon the logic circuit. The output becomes the function of input after a finite time delay. The delay for each output will depend upon the individual gate delays and the number of gates used in the circuit.

- Combinational circuit consists of logic gates whose outputs at any time are determined by combining the values of applied inputs using logic operations and sequential circuits.

- In some circuits, the output at any instant of time depends upon the present inputs as well as past inputs and outputs, such circuits are known as *sequential circuits.* In this chapter, we will study combinational circuits and in the next subsequent chapter we will study sequential circuits.

(2.1)

"

- The most popular combinational circuits are Half Adder, Half Subtractor, Full Adder and Subtractor, Multiplexers, Demultiplexers, Decoders and Encoders.

2.2 Half Adder

- A digital computer can perform arithmetic operations. e.g. addition, subtraction, multiplication etc. To perform these operations some building blocks are necessary. One of the important building block is half adder.

- As we know the basic rules for addition i.e. $0 + 0 = 0$, $0 + 1 = 1$, $1 + 0 = 1$ and $1 + 1 = 10_2$, the logic circuit which adds two binary digit at a time is called **half adder**.

- The half adder accepts two binary digits on its input and produces two binary digits on its output i.e. a **sum** and **carry** bit as shown in Fig. 2.1.

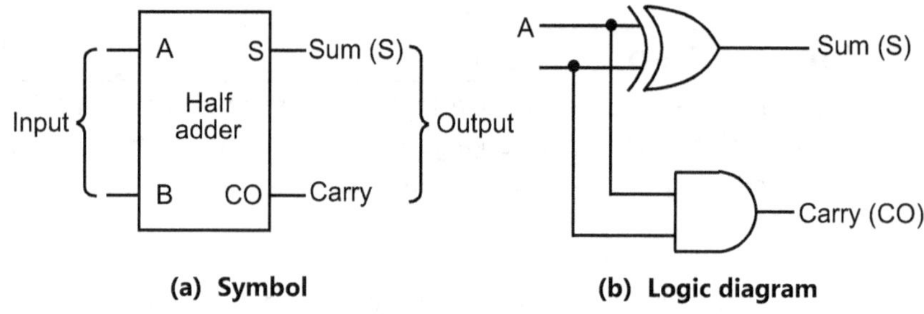

(a) Symbol　　　　　　　　　　(b) Logic diagram

Fig. 2.1

Table 2.1 : Truth table of half adder

Inputs		Outputs	
A	**B**	**Sum (S)**	**Carry (CO)**
0	0	0	0
0	1	1	0
1	0	1	0
1	1	0	1

From the truth table, we obtain the logical expression for sum and carry.

$$\text{Sum (S)} = \bar{A} B + A\bar{B}$$
$$= A \oplus B$$

and　　　　$\text{Carry (CO)} = AB$

- The realization of an half adder is shown in Fig. 2.1 (b), in which half adder is constructed with EX-OR and an AND gate. The output of AND gate is carry and output of the EX-OR gate is sum.

  Case I : Input A = 0 and B = 0. Both inputs of EX-OR gate are 0, so output of sum (S) is 0 and as both inputs of AND gate are 0 so output of carry (CO) is 0.

  Case II and III : When either input is 1 (i.e. A = 0 and B = 1 or A = 1 and B = 0) EX-OR output is 1, means sum (S) = 1. But still AND gate output (CO) = 0.

  Case IV : When A = B = 1 then again EX-OR output i.e. sum (S) = 0 and AND gate output i.e. carry (CO) = 1.

2.3 Half Subtractor

- Half subtractor is a logic circuit for the subtraction of B (subtrahend) from A (minuend). The symbol of half subtractor is shown in Fig. 2.2 (a). A half subtractor can subtract two bits and produces the difference (D) and borrow (BO).

- It can be implemented by using an EX-OR gate, AND gate and NOT gate. Input A is complemented by an inverter before giving it to the AND gate.

- Here the EX-OR gate produces the difference (D) and AND gate produces borrow (BO).

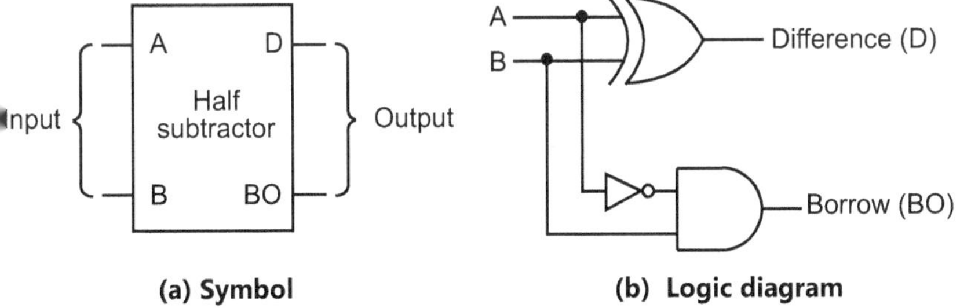

(a) Symbol   (b) Logic diagram

Fig. 2.2

Table 2.2 : Truth table of half subtractor

Inputs		Outputs	
A	**B**	**D**	**BO**
0	0	0	0
0	1	1	1
1	0	1	0
1	1	0	0

From the truth table, the logical expressions for D and BO are obtained as

$$D = \overline{A} B + A\overline{B}$$

$$= A \oplus B$$

$$BO = \overline{A} B$$

The realization of a half subtractor using gate is shown in Fig. 2.2.

- When A and B inputs both are 0 i.e. A = 0 and B = 0, EX-OR output is 0 i.e. difference D = 0 and AND gate output is also 0. i.e. BO = 0.

- When A = 0 and B = 1, 1 cannot be subtracted from 0, so one is borrowed and the difference will be (10 – 1 = 1) 1 i.e. D = 1 and BO = 1. On the other hand, when A = 1 and B = 0, the output of EX-OR gate will be 1 and AND gate output will be 0 i.e. D = 1 and BO = 0.

- When A = 1 and B = 1, the output of EX-OR gate is 0 i.e. difference D = 0 and the output of AND gate is 0 i.e. BO = 0.

2.4 Full Adder

- A half adder has only two inputs and there is no provision to add a carry coming from the lower order bit when multi-bit addition is performed.

- The full adder accepts three inputs and generates a sum output and a carry output.

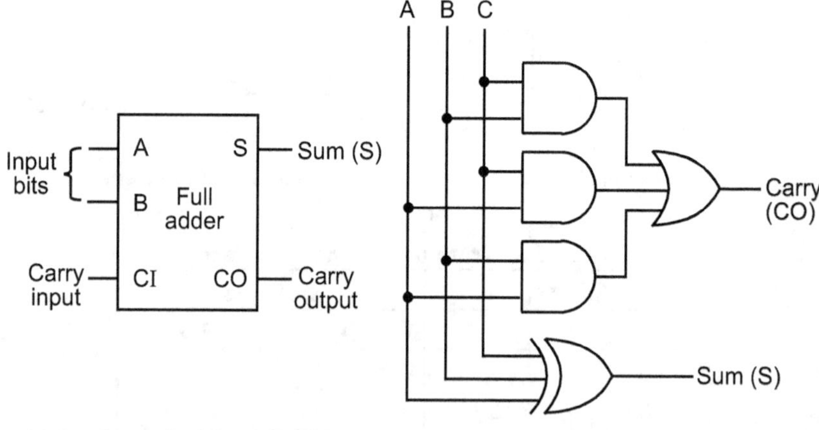

(a) Logic symbol for full adder   **(b) Logic diagram**

Fig. 2.3

Table 2.3 : Truth table of full adder

Inputs			Outputs	
A	**B**	**CI**	**Sum (S)**	**Carry (CO)**
0	0	0	0	0
0	0	1	1	0
0	1	0	1	0
0	1	1	0	1
1	0	0	1	0
1	0	1	0	1
1	1	0	0	1
1	1	1	1	1

- As full adder performs binary addition on 3-bits, we may write the logical expression for

  Sum $(S) = (A \oplus B) \oplus C$ and Carry $CO = AB + BC + AC$

  We may verify the truth table as follows :

  Case I : When $A = B = CI = 0$, sum $(S) = 0$ and $CO = 0$ because any one of the input of AND gate is 0, output becomes 0. i.e. $CO = 0$. EX-OR has output 0 if all inputs are 0 i.e. $S = 0$.

  Case II : When $A = B = 0$ and $CI = 1$, sum $= 1$ because single input of EX-OR is 1, output of all AND gates is 0, therefore, $CO = 0$.

- In this way, we can verify that all entries in the truth table are verified and we see that when two or more of A B CI inputs are high, carry CO is high and when odd number of A B CI drives input of EX-OR gate, it produces high output or sum $S = 1$.

- A full adder can also be implemented with two half adders and one OR gate as shown in Fig. 2.4. The second half adder gives the sum and OR gate gives the carry.

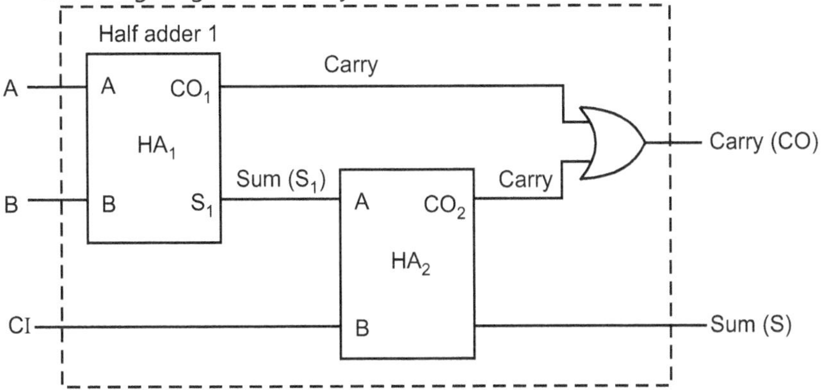

(a) Full adder logic diagram with two half adders

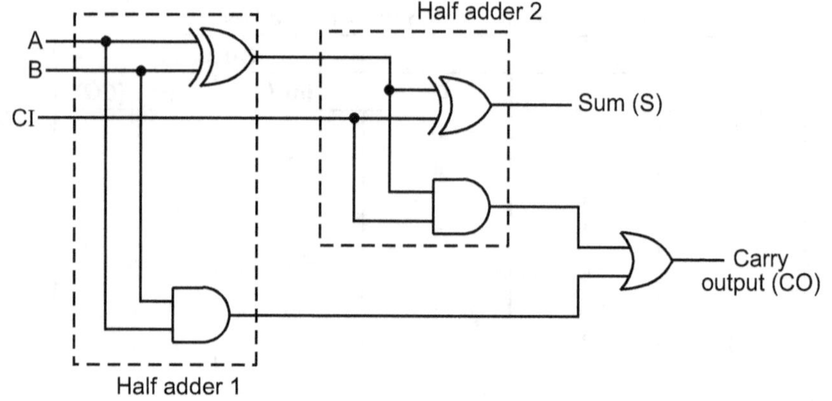

(b) Full adder using half adder

Fig. 2.4

2.5 Parallel Adder

- Full adder is capable for addition of 3-bits i.e. two one bit number and an input carry. In order to add binary numbers with more than one bit, additional full adders must be employed.

- When one binary number is added to another, each column generates a sum and 1 or 0 carry to the next higher column as illustrated with three bit numbers.

Carry from right column

```
  11              11
 111             111
 001             001
----            ----
1000            1000
```

↑ carry from third column becomes a sum bit

- To implement the addition of binary numbers with logic circuits, a full adder is required for each column, so for 3-bit numbers, three adders are needed, for 4-bit number, four adders are used and so on.

- The carry output of each adder is connected to the carry input of next higher order adder as shown in Fig. 2.5.

- A 3-bit parallel adder can be constructed with three full adders. These bit numbers can be added as,

C_3	C_2	C_1	C_0 (0)	Carries
	A_2	A_1	A_0	
	B_2	B_1	B_0	
	S_2	S_1	S_0	Sum

- There is no carry in for the first order C_0 (0), so first adder adds only A_1, B_0 and produces sum S_0 and carry C_1. The second adder adds A_1, B_1 and C_1, and produces sum S_1 and carry C_2. The third adder adds A_2, B_2 and C_2, and produces sum S_2 and carry C_3 . The output carry C_3 becomes most significant sum bit.

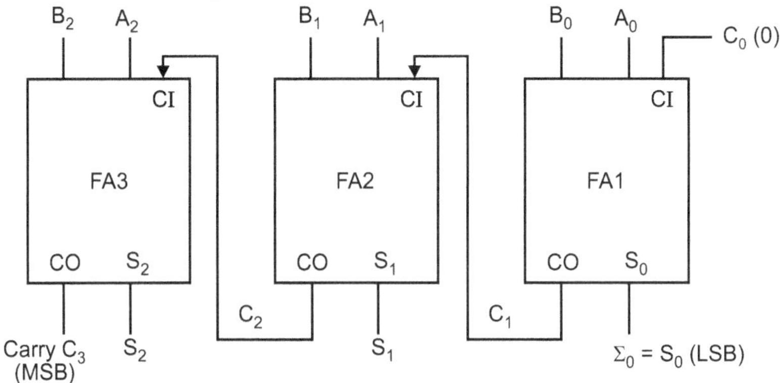

Fig. 2.5 : 3-bit parallel adder

Fig. 2.6 shows block diagram for 4-bit parallel adder.

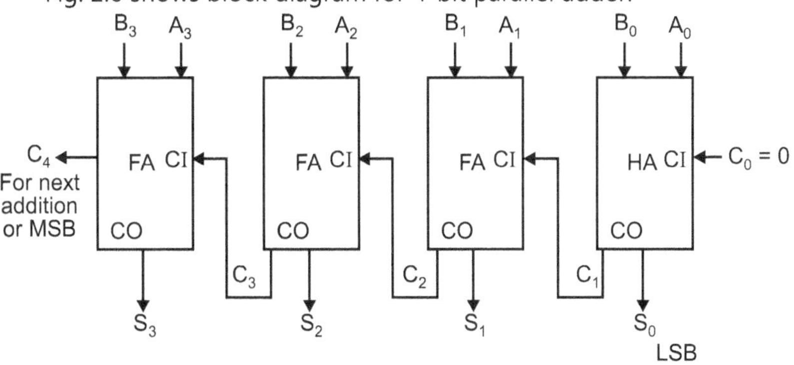

Fig. 2.6 : 4-bit parallel adder

- We can connect adders as shown in Fig. 2.6. To add group of two binary numbers, FA is full adder and HA is half adder. The binary numbers being added are $A_3 A_2 A_1 A_0$ and $B_3 B_2 B_1 B_0$.

$$\begin{array}{ccccc}
 & A_3 & A_2 & A_1 & A_0 \\
+ & B_3 & B_2 & B_1 & B_0 \\
\hline
\boxed{C_4} & S_3 & S_2 & S_1 & S_0 \\
\end{array}$$

$\uparrow$

carry for next addition

- Or, it may be written as $S_4\,S_3\,S_2\,S_1\,S_0$ also. The first column requires only half adder. For any column after the first, there may be a carry from the preceding column, therefore, we must use a full-adder for each column after the first column.

- For example, if we want to add decimal numbers 12 and 8, the binary equivalent of 12 is 1100 and for decimal 8 is 1000. The following Fig. 2.7 shows the binary adder with these inputs.

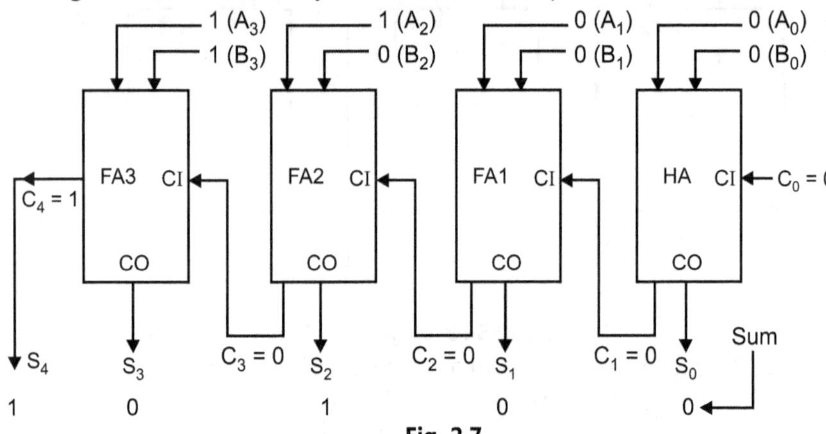

Fig. 2.7

- Starting from a half adder, we know the sum must be 0 with carry 0 as shown. The carry goes into first full adder FA1 which adds 0 + 0 + 0 i.e. $A_1 + B_1 + C_1$ to get a sum of 0 with carry 0. i.e. $S_1 = 0$, $C_2 = 0$.

- This carry $C_2 = 0$ goes into next full adder i.e. FA2, which adds 1 + 0 + 0 i.e. $A_2 + B_2 + C_2$ to get sum of 1 with carry 0. i.e. $S_2 = 1$ and $C_3 = 0$.

- The last full adder FA3 adds 1 + 1 + 0 i.e. $A_3 + B_3 + C_3$ to get sum of 0 and carry of 1. i.e. $S_3 = 0$ and $C_4 = 1$.

  The final output of the system is,

$$10100 = \boxed{C_4}\ S_3\,S_2\,S_1\,S_0$$

or

$$10100 = S_4\ S_3\,S_2\,S_1\,S_0$$

The decimal equivalent of this is

$$1\ 0\ 1\ 0\ 0$$

$$16\ \cancel{8}\ 4\ \cancel{2}\ \cancel{1} = 16 + 4 = 20$$

which is the correct decimal sum of 12 and 8. Therefore, the parallel binary adder of Fig. 2.7 gives the binary sum of two 4-bit numbers.

- The adder of Fig. 2.7 has limited capacity. The largest binary numbers that can be added are 1111 and 1111, so the maximum capacity is,

$$
\begin{array}{rr}
15 & 1111 \\
+\ 15 & +\ 1111 \\
\hline
30 & 11110
\end{array}
$$

- To increase the capacity, more full adders can be connected to the left end of the system. For instance, to add 5-bit numbers, we should connect one more full adder.

2.5.1 IC 7483

- IC 7483 is an MSI device that performs binary addition of two 4-bit numbers. Fig. 2.8 illustrates the block diagram.

- $A_3 A_2 A_1 A_0$ is one of the input binary numbers and $B_3 B_2 B_1 B_0$ is the other. The 7483 adds these binary numbers and delivers output of $S_3 S_2 S_1 S_0$, plus a carry of C_4.

  - For longer binary numbers, the bits can be processed in 4-bit nibbles. To add 20 bit numbers, five 7483s can be operated in parallel. Each 7483 is handling a 4-bit nibble. The C_4 carry of one 7483 becomes the C_0 input to the next stage.

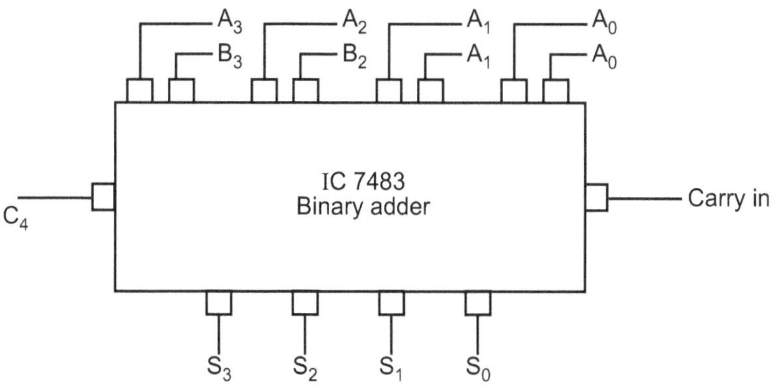

Fig. 2.8 : IC 7483 (4-bit binary adder)

2.6 Parallel 4-bit Binary Subtractor Full Subtractor

- Fig. 2.9 shows a circuit that subtracts $B_3 B_2 B_1 B_0$ from $A_3 A_2 A_1 A_0$. The first four inverters complement each B bit to get $\bar{B}_3 \bar{B}_2 \bar{B}_1 \bar{B}_0$, the 1's complement of $B_3\,B_2\,B_1\,B_0$. The full adders now add $A_3\,A_2\,A_1\,A_0,\ \overline{B}_3\,\overline{B}_2$

$\overline{B}_1\,\overline{B}_0$ and end around carry.

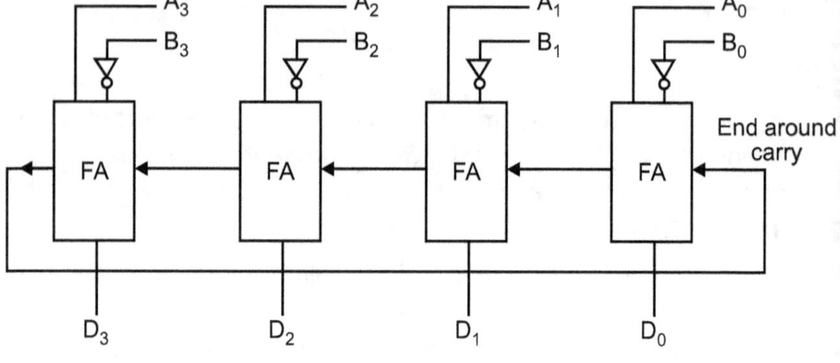

Fig. 2.9

- In this type of subtractor, 1's complement subtraction method is used. Instead of subtraction of number, we add the 1's complement of number. The last carry is then added to get the final answer. The last carry is called as end-around carry.

For example, subtract 8 from 12.

$$
\begin{array}{r r c c c c l}
12 & & 1 & 1 & 0 & 0 & = A_3\,A_2\,A_1\,A_0 \\
-\ 8 & & -\ 1 & 0 & 0 & 0 & = B_3\,B_2\,B_1\,B_0 \\
\hline
4 & & D_3 & D_2 & D_1 & D_0 &
\end{array}
$$

We have formed 1's complement of

$$B_3\,B_2\,B_1\,B_0 \;=\; 1000 \ \text{which is} \ \ \overline{B}_3\,\overline{B}_2\,\overline{B}_1\,\overline{B}_0 \;=\; 0111.$$

Now, add 1's complement i.e. $\overline{B}_3\,\overline{B}_2\,\overline{B}_1\,\overline{B}_0$ to $A_3\,A_2\,A_1\,A_0$.

$$
\therefore \qquad
\begin{array}{r}
1100 \\
+\ 0111 \\
\hline
\boxed{1}\ \ 0011 \\
\end{array}
$$

End around carry $\rightarrow$ + 1

$$0100$$

Result   0100 = 4

For above example, parallel 4-bit subtractor can be shown below.

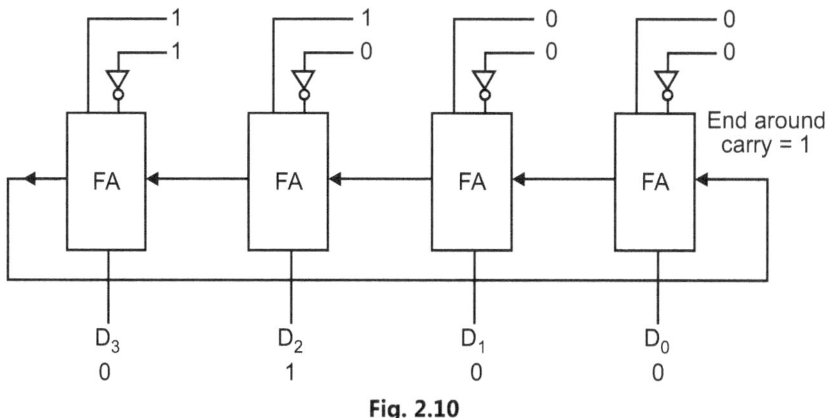

Fig. 2.10

2.7 Universal Adder/Subtractor (2's Complement Adder/Subtractor)

- This operation is based on 2's complement method which acts as an adder/subtractor, which can be illustrated as given in Fig. 2.11.

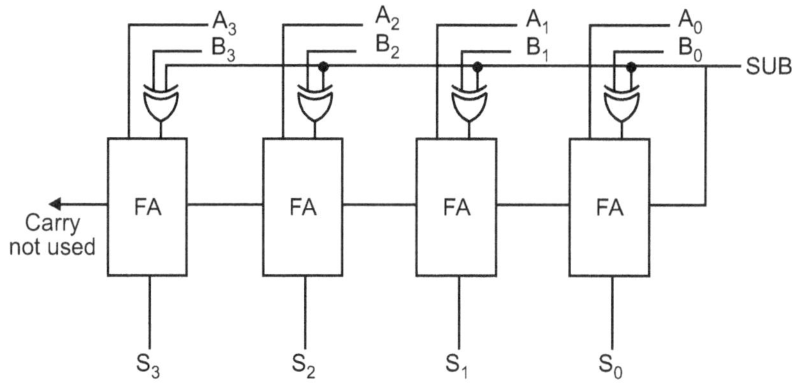

Fig. 2.11 : 2's complement adder/subtractor

- When SUB is low, the B bits pass through the controlled inverter to the full-adder. Therefore, the full-adders produce the sum of A and B. For example, A = 1001 and B = 0011, the decimal and binary sums are,

$$
\begin{array}{lll}
9 & 1001 & A_3\,A_2\,A_1\,A_0 \\
+\ 3 \ = & +\ 0011 & =\ B_3\,B_2\,B_1\,B_0 \\
\hline
12 & 1100 & S_3\,S_2\,S_1\,S_0
\end{array}
$$

- When SUB is high, the B bits are inverted before reaching full adders. Also, a high SUB adds a 1 to the first full-adder. This addition of 1 forms the 2's complement of B. In other words, the controlled inverter produces the 1's complement and adding 1 results in the 2's complement.

- Therefore the output of the full-adders is the difference of A and B. For instance, with the same values of A and B used in the preceding example, the decimal and binary sums when SUB is high, will be as given below,

$$
\begin{array}{cc}
9 & 1001 \\
-3 & +\ 1101 \ \rightarrow \ \text{2's complement of 3} \\
\hline
6 & 0110 \\
\end{array}
$$

2.8 Magnitude Comparator

- The comparators are used to compare the magnitudes of two quantities in order to determine the relationship of these quantities or we may say a comparator circuit determines whether two numbers are equal.

- The EX-OR gate is basic comparator because its output is 1 if its two input bits are not equal and output 0 if its inputs are equal.

Fig. 2.12 shows EX-OR as two bit comparator.

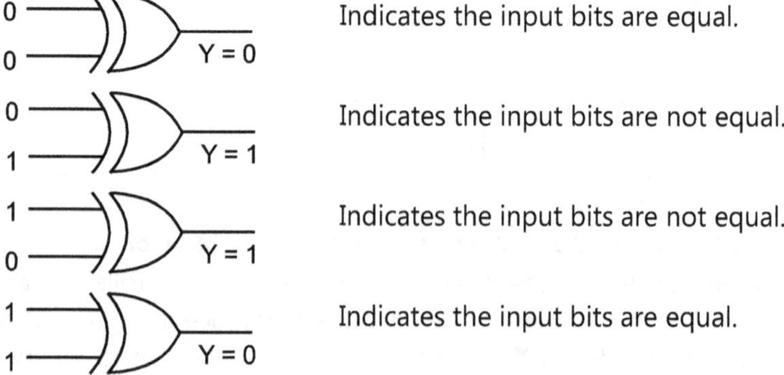

Fig. 2.12

- In order to compare binary number, containing two bits each, an additional EX-OR gate is necessary. The two least significant bits (LSBs) of the two numbers are compared by gate G_1, and the two most significant bits (MSBs) are compared by gate G_2 as shown in Fig. 2.13 which compares two bit binary numbers.

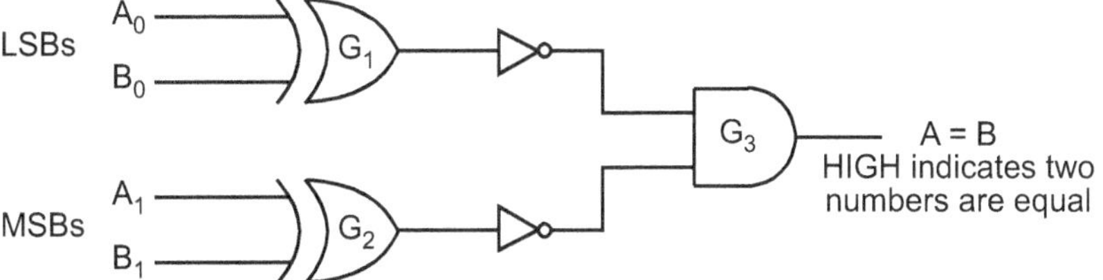

Fig. 2.13 : Logic diagram of 2-bit comparator

General format:

$$\text{Number A} = A_1 A_0$$

$$\text{Number B} = B_1 B_0$$

- If the two numbers are equal, their corresponding bits are same, output of gates G_1 and G_2 is 0, it is inverted by NOT gate, so the output of both inverters becomes 1. This 1 is applied to both inputs of AND gate G_3, so the output of AND gate G_3 i.e. 1 (HIGH). This high indicates two numbers are equal.

- The output of AND gate indicates equality (1) or inequality (0) of the two numbers.

  The exclusive-OR gate and inverter can be replaced by EX-OR gate.

2.8.1 Digital Lock using Magnitude Comparator

- The electronic locking system can be designed and constructed using simple circuit elements and implemented with logic gates. The magnitude comparator circuit can be used to construct digital lock. It can be used for crime and intrusion prevention. Like a conventional key, the electronic lock accepts a 2-digit code, compares it to the stored code in its memory via a comparator and switches ON or OFF any device connected to it if both codes match or mismatch.

- The digital lock will require keypad for data entry, a set of shift register for memory, EX-OR logic gates for comparing and a relay for switching.

- The device keypad for data entry can be achieved with a Binary Coded Decimal (BCD) Encoder IC (74C922), a set of shift register ICs (4035B) for memory and a set of 4001 quad NOR and 4070 quad EX-OR logic gates for comparing and a relay for switching.

- During testing, the result showed that the device switches ON or OFF any electronic device connected to it if both codes match or mismatch.

2.9 Multiplexers

- Multiplexer is one of the important combinational circuit which is widely used standard circuit in digital design. *Multiplexer* is 'a logic circuit that selects one out of several data inputs to a single output'.

- The input selection depends upon a set of select inputs. Multiplexer also means many into one, or it can also be regarded as single pole switch which selects one out of many input lines. Fig. 2.14 shows the block diagram of a multiplexer.

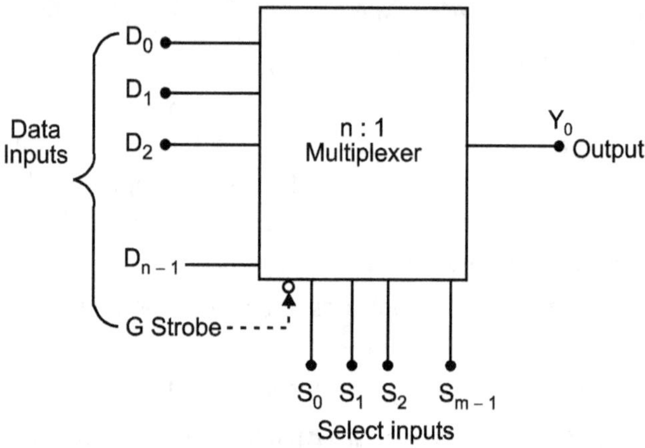

Fig. 2.14 : Block diagram of multiplexer

- In case of multiplexer, information present on either inputs is transferred on the output depending upon the logic levels of control signals. For n input multiplexers, m control lines are such that $2^m = n$.

- If all select lines are low, 1st input line gets selected whereas when all select lines are high, last input line gets selected and gets connected to the output line.

- Besides data input and select/control input, multiplexers usually have one more input known as *strobe* (or enable) *input* (G). Strobe is normally active low. Strobe input helps to enable or disable entire multiplexer circuit chip. It also helps in cascading the multiplexers.

- In its simplest form, multiplexer has two inputs and hence one select input and one output. Such circuit is known as 2 : 1 multiplexer.

2.9.1 2 : 1 Multiplexer

Fig. 2.15 (a) shows the block diagram and Fig. 2.15 (b) shows the truth table of 2 : 1 multiplexer.

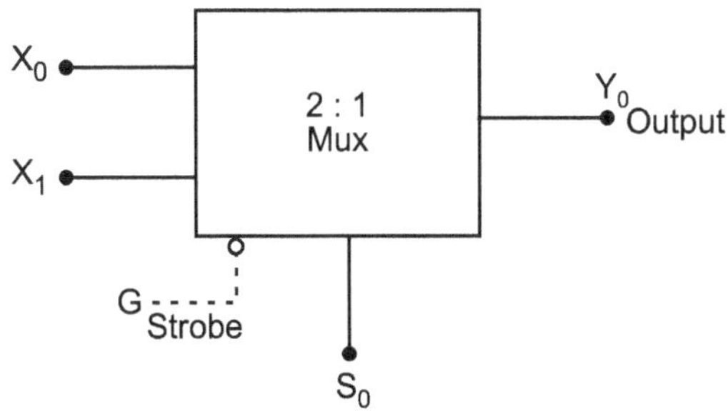

(a) Block diagram of 2 : 1 mux

Strobe	Select	Data	Input	Output	Remark
G	S_0	X_1	X_0	Y_0	
1	X	X	X	$-$	Disabled
0	0	X	0	0	} X_0 is
0	0	X	1	1	multiplexed
0	0	0	X	0	} X_1 is
0	0	1	X	1	multiplexed

(b) Truth table

Fig. 2.15

According to definition as shown by the truth table, when $S = 0$, $Y_0 = X_0$ and when $S = 1$, $Y_0 = X_1$. Fig. 2.16 shows the circuit diagram of 2 : 1 multiplexer.

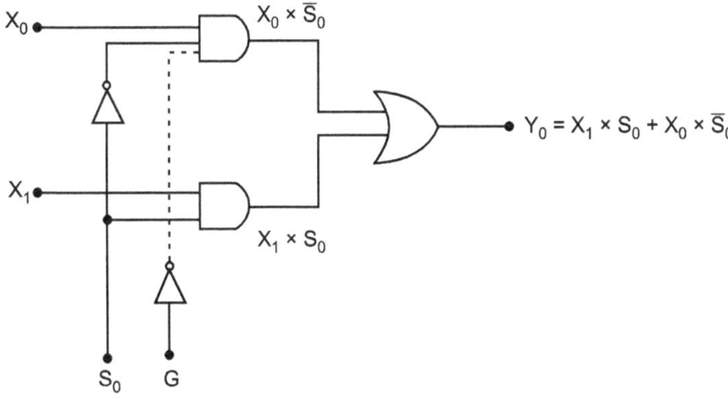

Fig. 2.16 : 2 : 1 mux using AND-OR gates

After learning logic gates, we know that NAND gates are universal gates. The 2 to 1 mux circuit shown in Fig. 2.16 can also be realized using only NAND gates as shown in Fig. 2.17.

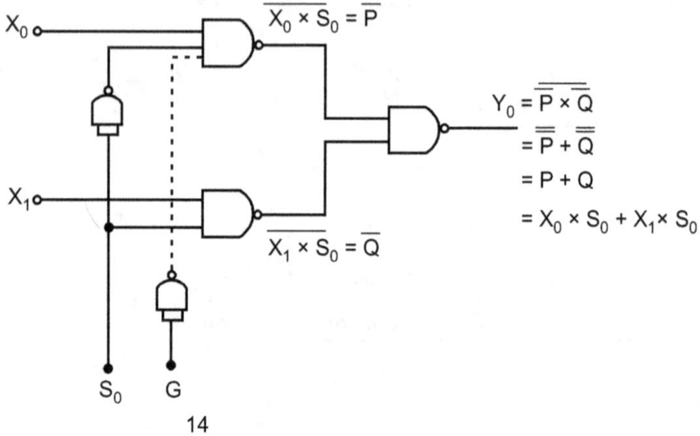

Fig. 2.17 : 2 : 1 mux using only NAND gates

In presence of strobe input, the output equation will be,

$$Y_0 = X_1 \cdot S_0 \cdot \overline{G} + X_0 \cdot \overline{S}_0 \cdot \overline{G}$$

when G = 1, Y_0 = 0
when G = 0, Case (i)   S = 0, Y_0 = X_0
                    Case (ii) S = 1 , Y_0 = X_1

2.9.2 4 : 1 Multiplexer

If the multiplexer has 4 inputs then it is known as 4 : 1 mux. To select one out of 4 inputs 2 select lines are required. Fig. 2.18 shows the circuit of 4 : 1 mux.

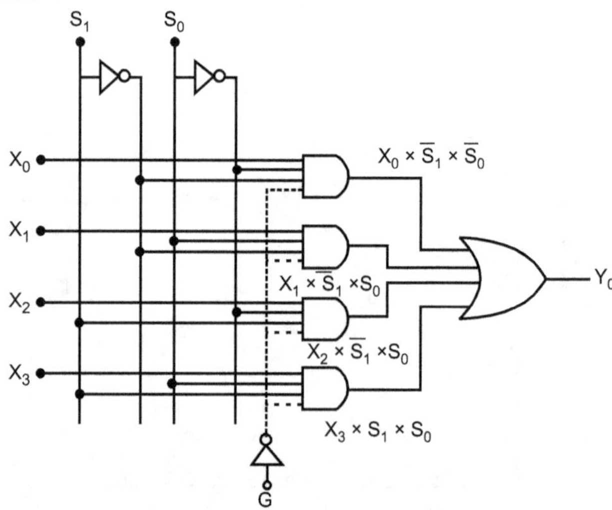

Fig. 2.18 : 4 : 1 mux using AND-OR gates

$$Y_0 = X_3 \cdot S_1 \cdot S_0 + X_2 \cdot S_1 \cdot \overline{S}_0 + X_1 \cdot \overline{S}_1 \cdot S_0 + X_0 \cdot \overline{S}_1 \cdot \overline{S}_0$$

Construction :

- From Fig. 2.18, we see that the circuit uses AND, OR and NOT gates. The dashed line indicates strobe input.

- The four data inputs are connected at the one input of four AND gates. The other two inputs to the AND gate are the select inputs.

- The outputs of all AND gates are connected to the input of OR gate.

Working :

- When strobe is high one input of all AND gates is low, all the AND gates are disabled and the output of the circuit is low.

- When the strobe is low and select inputs S_0 and S_1 are also low, upper G_0 and gate is enabled and all other AND gates have atleast one input low.

- So the data input X_0 gets coupled to the input of OR gate and hence output Y follows data input X_0. Similarly, when $S_0 = 1$ and $S_1 = 0$, Y_0 follows X_1 and so on. The truth table below shows the 4 : 1 mux.

Strobe	Select		Data input				Output
G	S_1	S_0	X_3	X_2	X_1	X_0	
1	X	X	X	X	X	X	0
0	0	0	X	X	X	X_0	X_0
0	0	1	X	X	X_1	X	X_1
0	1	0	X	X_2	X	X	X_2
0	1	1	X_3	X	X	X	X_3

Truth table

X = irrelevant

$X_{i \ (i = 0 \ to \ 3)}$ = 0 or 1

- Similar to 2 : 1 and 4 : 1 multiplexers, we can also construct 8 : 1, 16 : 1, 32 : 1 multiplexers. For 8 : 1 mux there are 8 inputs and to select 8 inputs there will be 3 select inputs. For 16 : 1 mux there will be 4 select inputs and for 32 : 1 mux there will be 5 select inputs.

- 8 : 1, 16 : 1 or 32 : 1 mux can be obtained using AND, OR and NOT gates or only using NAND gates. It is also found that 8 : 1 mux can be obtained using two 4 : 1 mux or 16 : 1 mux can be obtained using 8 : 1 mux. Fig. 2.19 shows the logic diagram of 8 : 1 mux using 4 : 1 mux and OR gate.

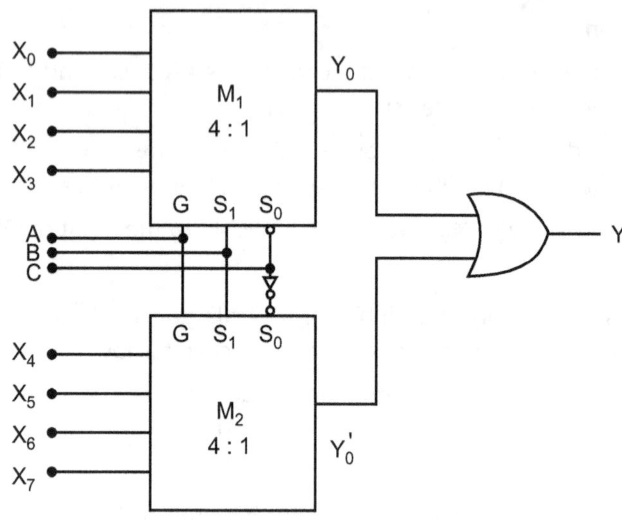

Fig. 2.19 : 8 : 1 mux using two 4 : 1 mux

- In the Fig. 2.19 there are 8 inputs X_0 to X_7 and three select inputs A, B and C. The X_0 to X_3 are inputs to M_1, X_4 to X_7 are inputs to M_2.

- The outputs of M_1 and M_2 are Y_0 and Y_0' respectively which are ORed together to get the final output Y. For this select input A is connected directly to strobe input of M_1 and is inverted and connected to strobe input of M_2, so that when A is zero, M_1 is selected and when A is 1, M_2 is selected.

- When C B A are all zero, Y_0 follows X_0 according to the truth table of 4 : 1 mux and at that time M_2 is not selected and hence Y follows X_0.

  Similarly, when C B A are all one, M_2 is selected. Y_0' follows X_7 and finally Y becomes X_7. Thus, depending upon select input Y follows X_0 to X_7.

- Similarly, using 8 : 1 we can obtain 16 : 1 and using 16 : 1 we can obtain 32 : 1 mux. Since 16 : 1 multiplexer IC is the largest available IC, such circuits are known as multiplexer tree.

2.9.3 4 : 1 Multiplexer using NAND Gates Only

Working :

- Multiplexers are the circuits having many input and only one output. 4 : 1 mux will have 4 input and 1 output. Such 4 : 1 mux circuit can be buit using AND-OR gates or using only NAND gates. Fig. 2.20 shows the circuit diagram of 4 : 1 mux using only NAND gates.

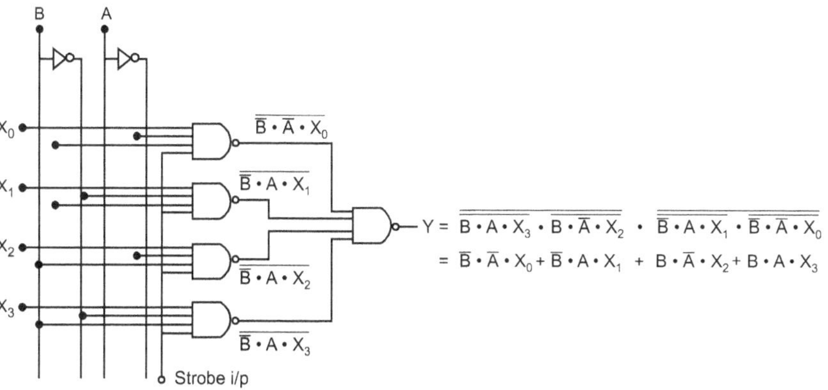

Fig. 2.20 : 4 : 1 multiplexer using only NAND gates

The use of strobe input is to enable or disable the entire multiplexer circuit.

2.9.4 Applications of Multiplexers

- Multiplexers find wide applications in the combinational system design because, if we use mux, simplification of logic expression is not required, it reduces the number of ICs which in turn saves PCB area, interconnections, propagation delays, power dissipation, component cost etc. Thus, if we use mux, logic design is simplified.

    1.   **Data routing :** The desired digital data from multiple sources can be routed at the destination using multiplexers under the control of data selection input.

    2.   **Time Division Multiplexing (TDM) :** The TDM technique is used in communication. Whenever there are many inputs trying to use the same transmission line to send its information to the destination. The data is transmitted using TDM. In this case, each input gets a time slot to send its information on the channel. Here select lines are used to select a particular input. The select lines are driven by a counter circuit and the inputs are selected periodically to load the signal on the channel.

- X-Y pattern detector, switch setting comparator and mux as the function generator are few more applications of multiplexer.

2.10 Demultiplexer

- A demultiplexer is a logic circuit which takes data from one line and routes it on one of many output lines under the control of select or control input. Demultiplexer means one into many and performs the reverse operation of multiplexer. Fig. 2.21 shows the block diagram of demultiplexer.

- In case of demux, the information present on the input is transmitted on one of the outputs, the select input determines to which output the data input will be transmitted. In order to select n outputs, m select inputs are required such that $n = 2^m$.

- As the demultiplexer takes data from one input line and routes it over 2^m output lines hence, demux are in the form 1 : 2 demux or 1 to 4 or 1 to 8 or 1 to 16.

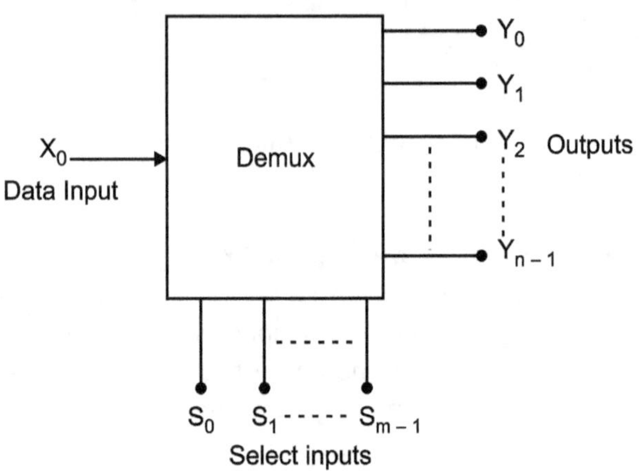

Fig. 2.21 : Block diagram of demultiplexer

2.10.1 1 : 2 Demultiplexer

- This configuration has single data input, one select or control and two outputs. The logic diagram is as shown in Fig. 2.22.

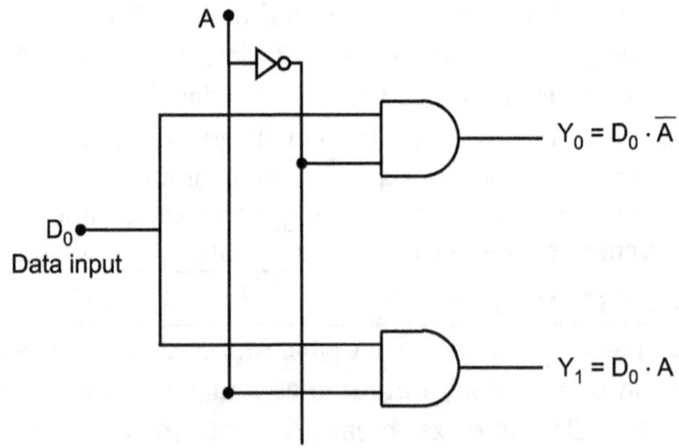

Fig. 2.22 : 1 : 2 demultiplexer

- When select is 0, data input is transferred on Y_0 whereas, when select is 1, data input is transferred to Y_1. The truth table of 1 : 2 demux is as given below.

Select	Outputs	
A	Y_0	Y_1
0	D_0	0
1	0	D_0

Truth table

Thus, data is transferred on Y_0 or Y_1 depending upon select or control input.

2.10.2 1 : 4 Demultiplexer

- Fig. 2.23 shows the circuit of 1 : 4 demultiplexer. To transfer data on 4 outputs, 2 select terminals are required. The circuit is implemented using AND and NOT gates.

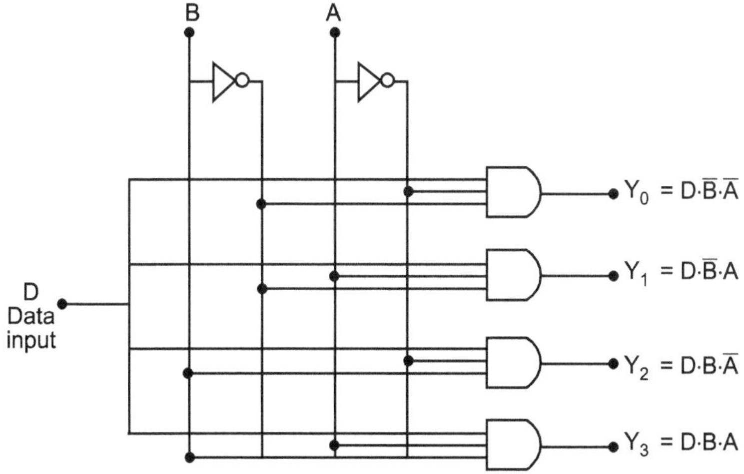

Fig. 2.23 : 1 : 4 demux

- From Fig. 2.23, boolean equation of Y_0 is $Y_0 = D \cdot \overline{B} \cdot \overline{A}$. From this equation, we see that $Y_0 = D$ only when B = 0, A = 0, i.e. D is transferred to Y_0 only when both selects are low. When B = 0 and A = 1, D is transferred on Y_1 and so on. The truth table of 1 to 4 demux is given below.

Select Inputs		Outputs			
B	A	Y_0	Y_1	Y_2	Y_3
0	0	D	0	0	0
0	1	0	D	0	0
1	0	0	0	D	0
1	1	0	0	0	D

Truth table

- Similarly we can also build 1 : 8 mux which has 3 select inputs and 8 outputs require eight 4 inputs AND gates and three NOT gates. Implementation of 1 to 16 line demux requires sixteen 5 inputs AND gates and four NOT gates.

- Even though demultiplexers can be constructed using discrete logic circuit, MSI chips are available containing different configurations of demultiplexers and multiplexers.

2.10.3 Applications of Demultiplexers

- Demultiplexers are combinational circuits whose action is controlled by select inputs that route input data to one of many outputs. Demultiplexers are useful in various applications.

    1. **Data demultiplexing :** To send single information to multiple destination, demux can be used. For example, an university can send same data to different affiliated colleges using demultiplexers.

    2. **Clock demultiplexing :** For functioning of digital circuits, control in the form of clock is always needed instead of using different clocks in different subsystems. A single clock can be used and can be routed to different subsystems using select input of demultiplexers.

    3. **Demultiplexer as a decoder :** Demultiplexers can be used as decoders. Decoders will be studied in next article but we can say that if data input is absent, demultiplexers get converted into decoders. Decoders are useful for memory addressing.

- Besides this, demultiplexers are also useful for switch encoding, for phase clock generation and for function generation.

2.11 Encoders

- Decimal code is the most widely used code in our day-to-day life. But the data expressed in decimal code is not accepted by digital system and hence it is necessary to convert a decimal code into a suitable code acceptable to the digital system, an encoder performs such task.
- An encoder is a circuit which converts the data expressed in a given more general code into the desired code.

2.11.1 Decimal to BCD Encoder

- This type of encoder has ten inputs, one for each decimal digit and four outputs corresponding to the BCD code. Fig. 2.24 shows the block diagram of decimal to BCD encoder.

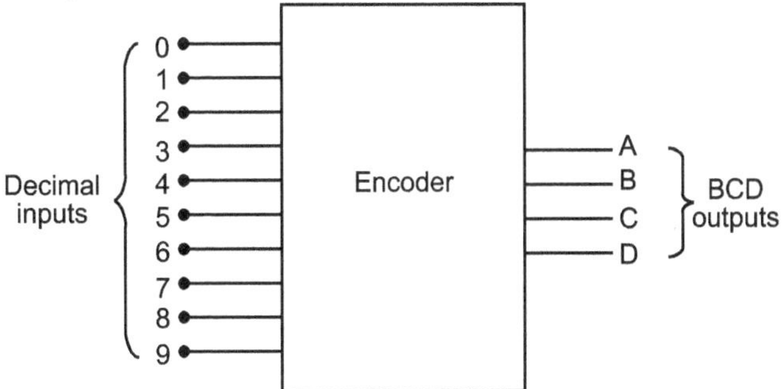

Fig. 2.24 : Block diagram of decimal to BCD encoder

Table 2.4 below gives BCD code corresponding to decimal digit.

Table 2.4

Decimal digit	BCD code			
	D	C	B	A
0	0	0	0	0
1	0	0	0	1
2	0	0	1	0
3	0	0	1	1
4	0	1	0	0
5	0	1	0	1
6	0	1	1	0
7	0	1	1	1
8	1	0	0	0
9	1	0	0	1

- The BCD code for each decimal digit is as shown in the Table 2.4. From table we can determine the relationship between each BCD bit and decimal digit. e.g. A is 1 for digits 1, 3, 5, 7, 9 i.e. the logical expression for A is,

$$A = 1 + 3 + 5 + 7 + 9$$

B is 1 for decimal digits 2, 3, 6, 7 and hence the expression for B is,

$$B = 2 + 3 + 6 + 7$$

Similarly, expression for C is,

$$C = 4 + 5 + 6 + 7$$

and that for D is $D = 8 + 9$

- Now, we can implement the logic circuit required for encoding each decimal digit to a BCD by using above logical expression. It simply consists of ORing the appropriate decimal digit input to form a BCD output.

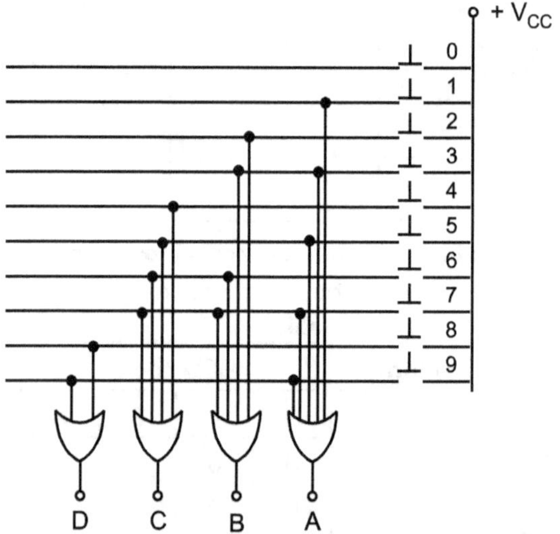

Fig. 2.25 : Decimal to BCD encoder using OR gates

- When any one button corresponding to decimal digit is pushed, the corresponding line becomes high. If button 5 is pressed, input line 5 becomes high, assuming all other inputs low, one input to OR gates A and C is high which produces a high on output A and C and low on output B and D which is BCD code 0101 for decimal digit 5. Similarly, when button 9 is pressed, the code generated will be,

$$DCBA = 1001$$

Thus, the circuit gives BCD code corresponding to decimal digit input.

- The circuit for decimal to BCD encoder can also be constructed using NAND gates, in which the input is active low and the output is active high for the circuit shown above.

- However, if more than one button is pressed simultaneously an erroneous code is generated by the circuit, this problem is solved by priority encoders.

2.11.2 Priority Encoders

- The priority encoder includes the necessary logic which is needed to ensure that when two or more inputs are activated, only the code which corresponds to highest number input is generated at the output.

- For example, if 5 and 7 inputs are made high simultaneously, the circuit will generate a code of 7 at the output.

- IC 74147 is decimal to BCD priority encoder. The block diagram of IC 74147 is shown in Fig. 2.26.

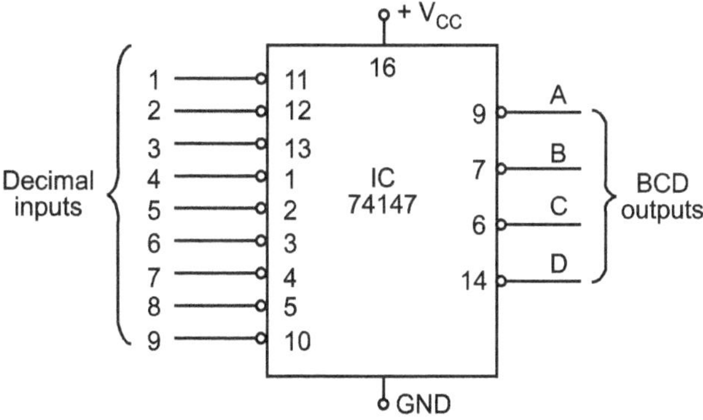

Fig. 2.26 : Block diagram of IC 74147

- IC 74147 has active low inputs and outputs and hence active low decimal input gets converted to complemented BCD outputs. IC 74147 is known as priority encoder as it gives priority to the highest number input.

2.11.3 Hexadecimal to Binary Encoder

- Hexadecimal to binary encoder has sixteen inputs 0 through F and four binary outputs. The logic circuit for hex to binary can be constructed on similar line of decimal to BCD.

Table 2.5 shows the binary equivalent of hex.

Table 2.5

Hex digit	Binary output			
	D	C	B	A
0	0	0	0	0
1	0	0	0	1
2	0	0	1	0
3	0	0	1	1
4	0	1	0	0
5	0	1	0	1
6	0	1	1	0
7	0	1	1	1
8	1	0	0	0
9	1	0	0	1
A	1	0	1	0
B	1	0	1	1
C	1	1	0	0
D	1	1	0	1
E	1	1	1	0
F	1	1	1	1

From Table 2.5, we see that the equation for A output is

$$A = 1 + 3 + 5 + 7 + 9 + B + D + F$$
$$B = 2 + 3 + 6 + 7 + A + B + E + F$$
$$C = 4 + 5 + 6 + 7 + C + D + E + F$$
$$D = 8 + 9 + A + B + C + D + E + F$$

By using 4, 8 input OR gates, we can construct hex to binary encoder. There is no standard IC for this encoder.

2.11.4 Keyboard Encoder

- To feed data to the computers and digital systems, keyboards are used. Typically keyboards are either hexadecimal or ASCII keyboard.

- In case of electronic calculator, there are ten switches corresponding to digit 0 through 9. When any key corresponding to particular digit is pressed, the internal circuitry should understand which key has pressed.

- We need a keyboard encoder which will generate a code corresponding to key pressed. Fig. 2.27 shows the simple keyboard encoder for 10 keys using IC 74147 as priority encoder.

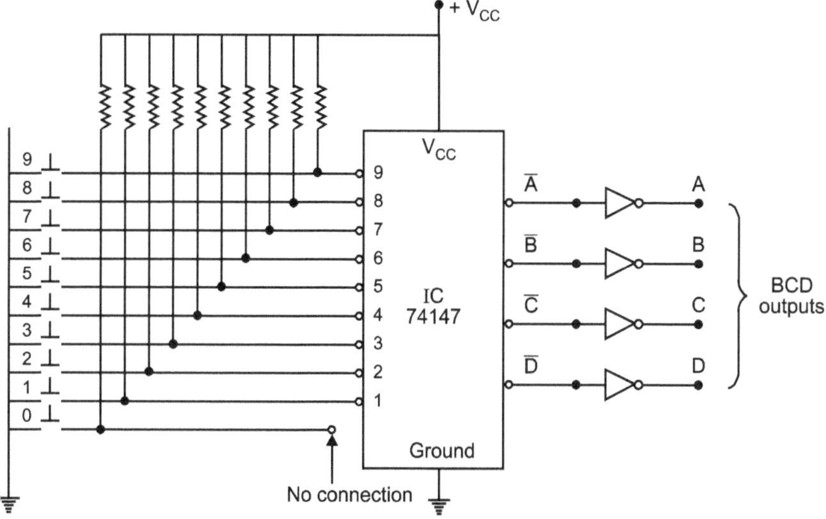

Fig. 2.27 : Keyboard encoder

- Notice that there are ten push-button switches each with pull-up resistor to $+V_{CC}$. The pull up resistor makes the line high when a switch is not depressed. When a switch is depressed the line gets connected to ground and a low is applied to the corresponding encoder input.

- The zero key is not connected because the BCD output represents zero when none of the other keys are depressed. When key corresponding to digit 9 is depressed as we have seen in the action of 74147, the IC generates the BCD code of 9 in complement form. It can be converted to proper BCD by using inverter at the output of IC as shown in Fig. 2.27.

2.11.5 3 × 4 Matrix Keyboard Encoder

- Fig. 2.28 shows 3 × 4 matrix keyboard. To detect a pressed key, initially all the rows are grounded by manually connecting them to zero. Then columns are read. If the data, read from the columns is $D_3 D_2 D_1 D_0 = 1111$, no key has been pressed.

- If one of the column bit has zero, this means that a key press has occurred.

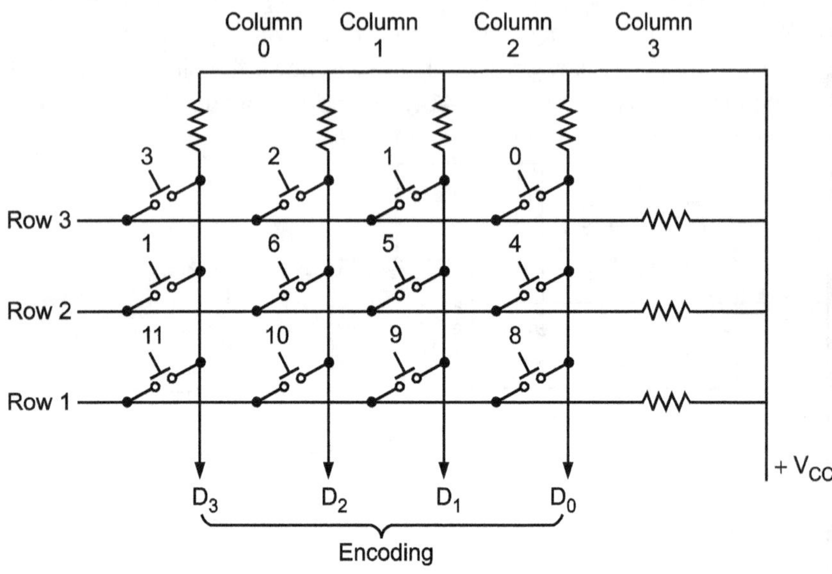

Fig. 2.28 : 3 × 4 matrix keyboard encoder

- For example, (i) If $D_3 D_2 D_1 D_0$ = 1101.

   This means that key in the D_1 column i.e. key no. 1, 5 or 9 has been pressed.

   (ii) If $D_3 D_2 D_1 D_0$ = 1110, we can identify definitely the key in the column D_0. i.e. key 0, 4 or 8 has been pressed.

2.12 Decoders

- Decoding is a process in which the information present in one code is obtained back into the desired code e.g. the binary output of a digital circuit is converted into decimal system by using decoders. Thus, decoders perform converse operation of encoders.

- A decoder is similar to a demultiplexer with exception that there is no data input. The only inputs are the control bits DCBA.

- A decoder is a special case of demultiplexer. A demultiplexer can pass a single input to any one of the outputs, depending upon the data status at control lines.

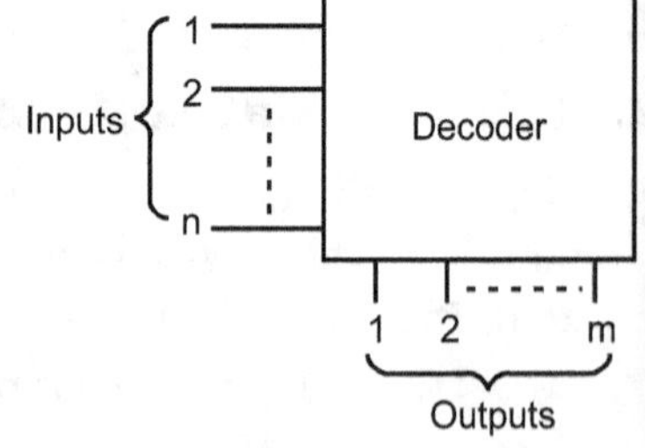

Fig. 2.29

- The only difference in decoder and demultiplexer is that, it does not has data input. The reverse process of encoding is known as decoding. A combinational circuit capable of decoding a data is called decoder. A decoder accepts a set of binary numbers and activates only the output corresponding to that input number.
- A decoder has 'n' inputs and 'm' outputs, for 'n' inputs we can have 2^n input combinations. For each of these combinations only one of the 'm' outputs will be high, all other outputs will be low. Some decoders, do not utilize all of the 2^n possible inputs but only few of them. For example, a BCD-to-decimal decoder has 4-bit input code and ten output lines. Such decoder is designed in such a way that if unused codes are applied at inputs, none of the output will be activated.
- The decoders can be referred to in several ways. For example, a decoder with 3 input lines and 8 (2^3 = 8) output lines can be called a 3-line-to-8-line decoder. It can also be called as binary-to-octal decoder (converter), because it accepts a 3-bit binary data and activates only one of the eight (octal) outputs corresponding to that code. The another name for the same is 1-of-8 decoder as only 1 of the 8 outputs is activated at a time.

2.12.1 1-of-4 Decoder

- 1-of-4 decoder is also called as 2-time to 4-line decoder as it has 2 input lines and 4 (2^2 = 4) output lines. The logic circuit is similar to 1-to-4 demultiplexer except the data input line is not there. The Fig. 2.30 shows the logic diagram of 1-of-4 decoder.

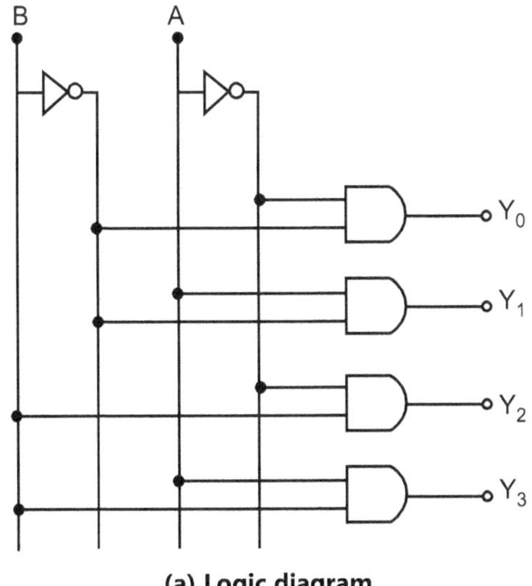

(a) Logic diagram

Inputs		Outputs			
B	**A**	Y_0	Y_1	Y_2	Y_3
0	0	1	0	0	0
0	1	0	1	0	0
1	0	0	0	1	0
1	1	0	0	0	1

(b) Truth table

Fig. 2.30 : 1-of 4 decoder

- When A = 0 and B = 0, the inputs to the first AND gate will be 1, hence the output will be 1. Output of other gates will be 0. Similarly, the inputs A = 1 and B = 1 make the output of fourth gate 1 and other gates 0. Thus, for any input combinations only one output will be high making other outputs low. IC 24139 is a 1-of-4 decoder.

- The IC 74154 is a decoder/demultiplexer since it can be used either as decoder or demultiplexer. It has active low outputs. The same IC can be used as decoder by grounding data and strobe input.

2.12.2 3 to 8 Decoder

- In case of 3 to 8 decoder there are three inputs and eight outputs.
- Any one of the eight output is active at a time.
- The circuit is also known as 1 of 8.
- Truth table of 3 to 8 decoder is as shown in Fig. 2.31.

Inputs			Output							
C	**B**	**A**	Y_7	Y_6	Y_5	Y_4	Y_3	Y_2	Y_1	Y_0
0	0	0	0	0	0	0	0	0	0	1
0	0	1	0	0	0	0	0	0	1	0
0	1	0	0	0	0	0	0	1	0	0
0	1	1	0	0	0	0	1	0	0	0
1	0	0	0	0	0	1	0	0	0	0
1	0	1	0	0	1	0	0	0	0	0
1	1	0	0	1	0	0	0	0	0	0
1	1	1	1	0	0	0	0	0	0	0

Fig. 2.31 : Truth table for 3 to 8 decoder

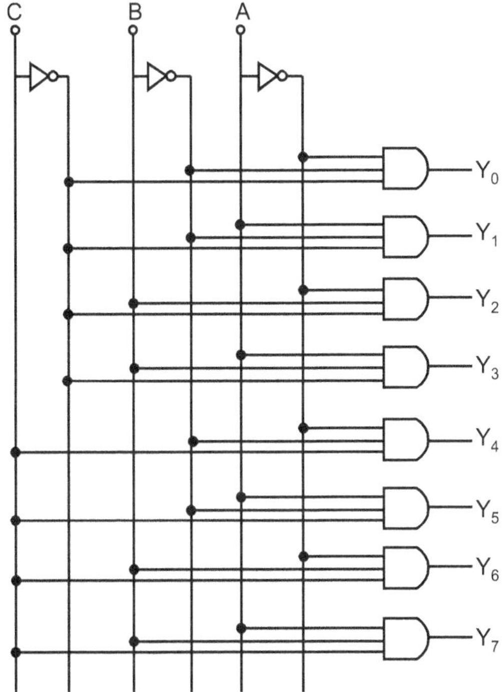

Fig. 2.32 : 3 to 8 decoder

2.12.3 BCD to Seven Segment Decoder

- In order to visualise decimal digit, seven segment displays are used. The seven segment display contains 7 LEDs arranged as shown in Fig. 2.33.

Fig. 2.33 : Seven segment display arrangement

Fig. 2.34 : Active segment for some digits

- The proper segments are activated to display the desired digit. Here seven segments are obtained by using seven LEDs. LEDs are basically P-N junction diode. When it is forward biased, free electrons recombine with holes near the junction as the free electrons fall from higher energy level to a lower level. They give difference in energy in the form of light.

- By using different material LEDs that emit different colours of light various colours can be obtained. As shown in Fig. 2.33, 7 LEDs are labelled a through g. In order to display particular digit, proper LED segments are forward biased e.g. to display 0 we need to forward bias segments a, b, c, d, e, f.
- In order to bias LEDs with minimum number of connections there are two arrangements or configurations of LEDs.
  1. **Common anode display :** In common anode type of display, the anodes of all LEDs are connected together and are given to the supply V_{CC}, the cathodes are left open as shown in Fig. 2.35.

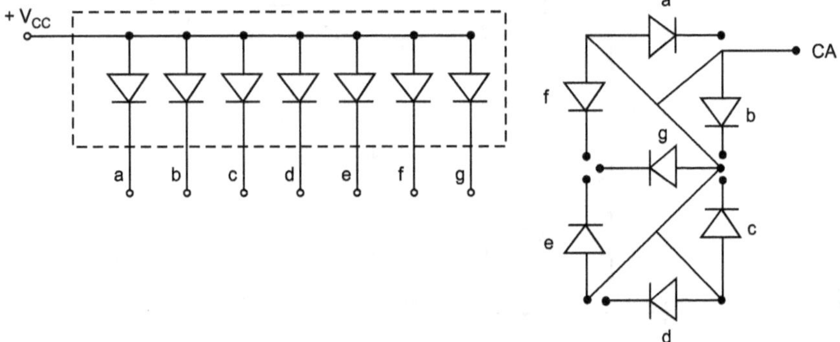

Fig. 2.35 : Arrangement of LEDs in CA display

  2. **Common cathode display :** In common cathode type of display, cathodes of all segments are connected together and finally tied to ground. Anodes are left open as shown in Fig. 2.36. In order to forward bias the segments only anode needs to be connected to V_{CC}.

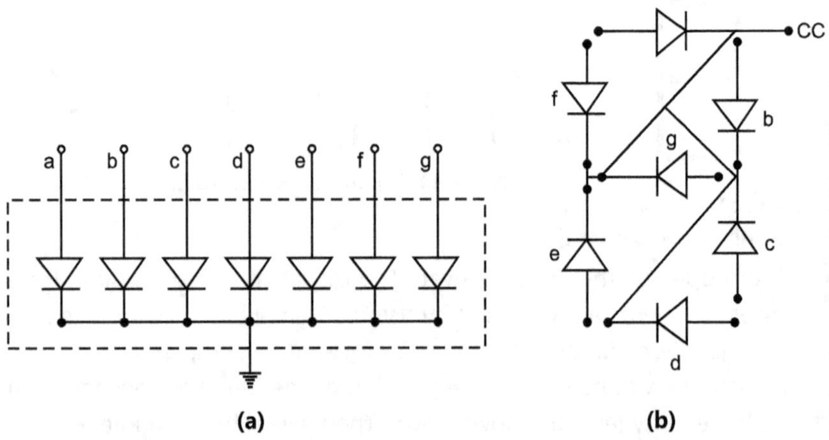

(a)   (b)

Fig. 2.36 : Arrangement of LEDs in CC display

- Display devices are used to display decimal digit equivalent to BCD. For this purpose, the data which is in BCD format has to be changed suitably.

- The circuit arrangement required for this is BCD to 7-segment decoder. The block diagram of BCD to 7-segment decoder is shown in Fig. 2.37 (a) and 7-segment LED display unit is shown in Fig. 2.37 (b).

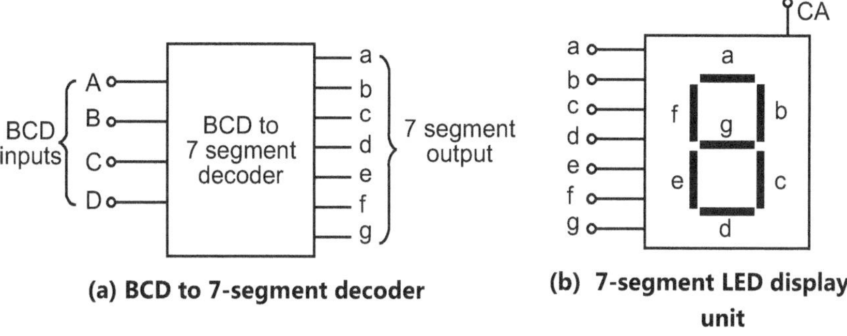

(a) BCD to 7-segment decoder

(b) 7-segment LED display unit

Fig. 2.37

- The BCD to 7-segment decoder will activate the outputs depending upon BCD inputs e.g. if BCD input is 0, all outputs except g are activated, which in turn visualizes decimal 0.

- A seven segment decoder/driver is an IC decoder that can be used to drive a seven segment indicator. There are two types of decoders/drivers, corresponding to common anode (7446/47) and common cathode (7448) display.

Fig. 2.38 shows 7446 driving a CA display.

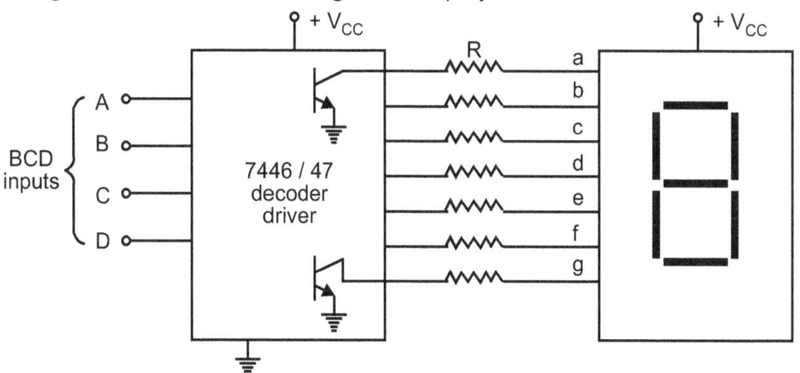

Fig. 2.38 : 7446 driving a CA display

- The logic circuit inside the 7446 converts the BCD input to the required output e.g. if BCD input is 0010, the internal logic of the IC 7446 will force LED a, b, g, e and d to conduct because the corresponding transistors go into saturation as a result digit 2 will appear on the seven segment indicator.
- When LED segment is forward biased, excessive current flowing through the junction may damage the junction (typical LED current is 50 mA). In order to limit the current through LED a current limiting resistor has to be connected in series with LED segment as shown in Fig. 2.38.
- The value of these resistors can be calculated by using Ohm's law. For appropriate brightness current through the segment is 15 mA and for 5 V supply, current limiting resistor is R = 5 V/15 mA = 330 Ω.
- IC 7448 is common cathode seven segment decoder/driver. IC 7448 does not require external current limiting resistor since it has its own current limiting resistor on the chip. Most of the decoder/driver IC chips have additional input/output terminals besides the BCD input and seven segment output.

2.13 Parity Generator and Checker

- Binary information is normally handled by a digital system in a group of bits which is called as *word*. A word always contains either even or odd numbers of 1s.
- A parity bit is attached to group of information bits in order to make the total number of 1s even or odd.
- Even parity means an n-bit input has an even number of 1s. For example, 101101 has even parity because it contains four 1s.
- Odd parity means an n-bit input has an odd number of 1s. For example, 110010 has odd parity because it contains three 1s.
  e.g. 1011 0010 1110   1000 even parity
       1111 0111 1000   0001 odd parity.
- The first binary number has even parity because it contains eight 1s, the second binary number has odd parity because it contains nine 1s.

2.13.1 Parity Checker

- As we know Exclusive-OR (XOR) gate has high output only when an odd number of input is high i.e. if single input is 1, output is 1. Therefore, XOR gates are ideal for checking the parity of binary number.
- An even parity input to an XOR gate produces a LOW output, while an odd parity input produces a high output.

Consider the following circuit having four input exclusive OR gate.

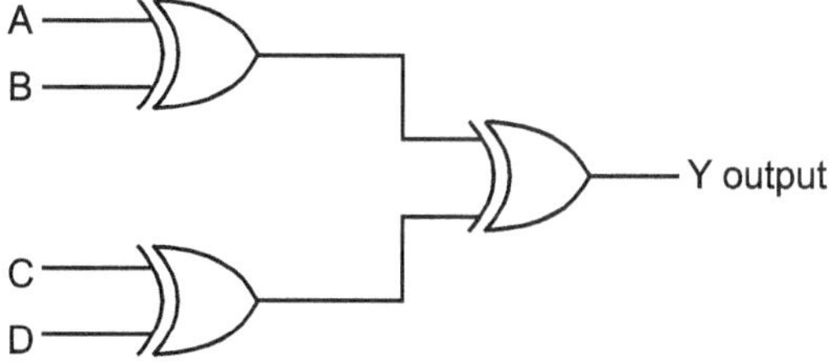

Fig. 2.39 : Four input exclusive OR gate

- A 4 input exclusive OR gate produces an output 1 when ABCD input has an odd number of 1 and it produces 0 when ABCD input has an even parity number.

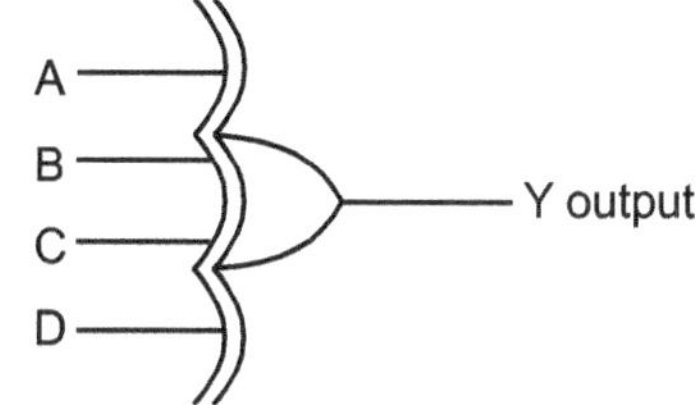

Fig. 2.40 : Abbreviated symbol for four input XOR

2.13.2 Parity Generator

- In a computer an instruction may be represented by a binary number. Instruction that tells computer to make arithmetic or logical operations such as addition, subtraction, multiplication and so on or the binary number may represent the data to be processed like a number, letter etc. In either case.
- we see that sometimes an extra bit is added to the original binary number to produce a new binary number with even or odd parity. For example, Fig. 2.41 shows 8 bit binary number which can be represented as, $D_0 D_1 D_2 D_3 \ D_4 D_5 D_6 D_7$.
- Suppose we apply input 01010000, then the number has even number of 1, means even parity, which means the exclusive-OR gate produces an output of 0 and $D_8 = 1$, because of inverter gate it becomes 1 and the final 9 bit output is 01010000 1. Note that, this is having odd parity.
- If we change the input number like 01011000 that is having number of 1 are odd means odd parity then output of XOR gate is 1 and inverter produces output $D_8 = 0$,   therefore, the number becomes 01011000 0, again the final output has odd parity.
- In both cases, 9-bit output number has odd parity so this circuit is called as *odd parity generator.*
- If we delete the inverter circuit (NOT gate is removed), this circuit can be applied for even parity generator.

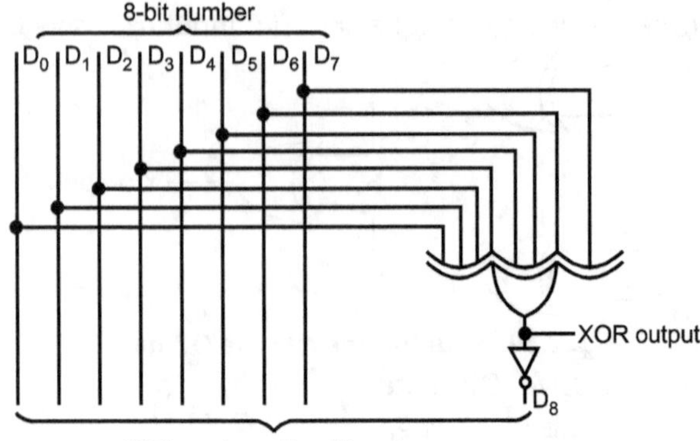

Fig. 2.41 : Odd parity generator

- Parity generator circuits are essential because of transients, noise and other disturbances, 1-bit error sometimes occurs when binary data is transmitted over telephone lines or other communication paths.

- One way to check for errors is to use an odd-parity generator at the transmitting end and an odd parity checker at the receiving end.

- If 1-bit error occurs in transmission, the received data will have odd parity. But if one of the transmitted bits is changed by noise or any other disturbance, the received data will have even parity. For example, we want to transmit data bits 10010001.

- With an odd parity generator as we have seen in Fig. 2.31, the data to be transmitted will be 10010001 0.

- This data can be sent over telephone lines to some destination. If no error occurs in transmission, the odd-parity checker at the receiving end will produce a high output, meaning the received number has odd parity.

- If a 1-bit error does creep into the transmitted data, the odd parity checker will have low output. It indicates that the received data is invalid. IC 74180 (14 pin) is available for parity generator checker.

Solved Examples

Example 2.1 : *Construct 4 : 1 multiplexer using two 2 : 1 multiplexers.*

Solution : Multiplexers are circuits having many inputs and only one output.

Multiplexers are combinational circuits and are obtained using logic gates.

Higher input multiplexers can be constructed using more number of lower number input multiplexers.

4 : 1 multiplexers can be constructed using two 2 : 1 multiplexers.

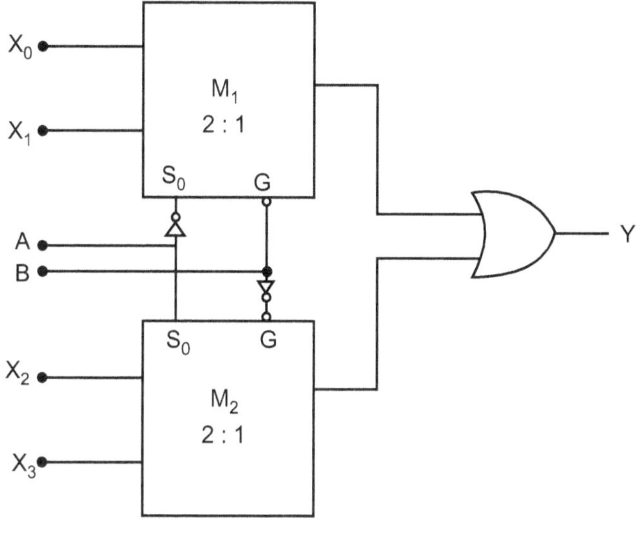

Fig. 2.42

Example 2.2 : *Draw the circuit to obtain 4 to 1 multiplexer using 2 to 1 multiplexer.*

Solution : We can obtain 4 to 1 multiplexer by using, two 2 to 1 multiplexers and one 2 input OR gate. Fig. 2.43 shows the necessary logic diagram.

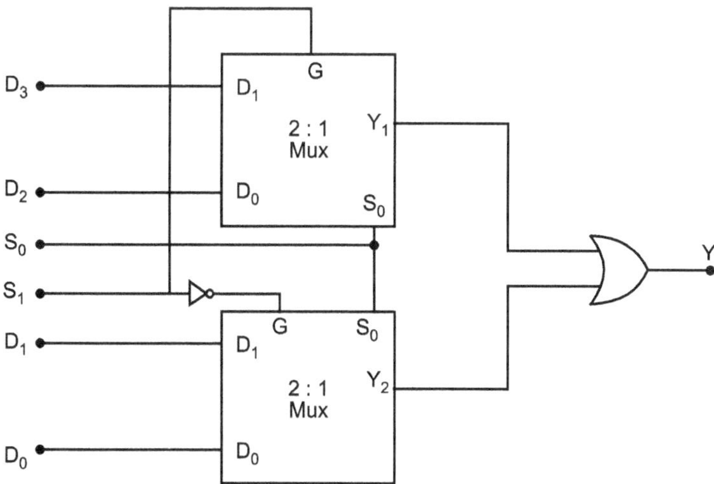

Fig. 2.43 : 4 to 1 mux using 2 to 1 mux

Here select inputs of both are combined together, whereas strobe is combined with an inverter as shown in Fig. 2.43. When S_0 and S_1 both selects are low, lower mux gets enabled and since S_0 is low Y_2 follows D_0, Y_1 is still low and hence one input to OR gate is D_0, other is 0, so Y becomes D_0 and so on. The truth table of the circuit is as shown below.

Select input		Output
S_1	S_0	Y
0	0	D_0
0	1	D_1
1	0	D_2
1	1	D_3

This is how we can obtain 4 to 1 multiplexer using 2 to 1 multiplexer.

Example 2.3 : *Implement the expression f (A, B, C) = $\sum m$ (0, 2, 4, 6) using a multiplexer.*

Solution : Here we need a circuit which gives the high output when the inputs 0, 2, 4 and 6 are high. Since there are three variables, therefore, a multiplexer with three select inputs is required. The circuit of 8 : 1 mux connected to implement the above expression is shown in Fig. 2.44.

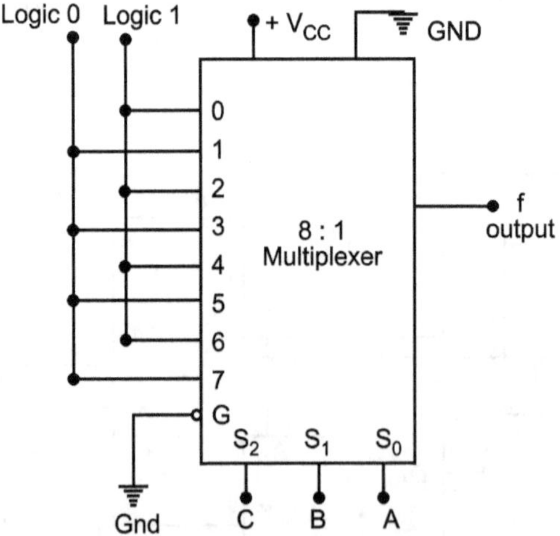

Fig. 2.44 : Implementation of logic expression

This implementation requires only one IC pack. We can use IC 74152 which is 8 : 1 multiplexer. However, it has output as inverted input, to obtain output same as input we may use inverter.

Think Over It

1. In case of combinational circuit output does not depends on previous input and outputs.
2. Full adder circuit can add any bit binary number.
3. Using 'n' select input we can multiplex 2^n inputs.
4. Encoders and decoders are must for processing decimal data.
5. BCD to seven segment decoders require for visualization of data.

Points to Remember

1. Digital circuits build using logic gates are classified into two categories : *combinational circuits* and *sequential circuits.*
2. The circuits whose output at any instant of time depends on input present at that instant of time and does not depends on any other factor are known as *combinational circuits.*
3. The circuits whose output at any instant of time depends upon the present inputs as well as past inputs and outputs are known as sequential circuits.
4. Multiplexer is a logic circuit that selects one out of several data inputs to a single output.
5. Multiplexer circuits are used in data selection, data routing, operation sequencing, parallel-to-serial conversion, waveform generation and logic function generation.
6. For n input multiplexers, m control lines are such that 2^m = n.
7. Strobe input helps to enable or disable entire multiplexer circuit chip. It also helps in cascading the multiplexers.
8. Multiplexers are used in the combinational system design because it reduces the number of ICs which in turn saves PCB area, interconnections, propagation delays, power dissipation, component cost etc.
9. A demultiplexer is a logic circuit which takes data from one line and routes it on one of many output lines under the control of select or control input.
10. The strobe line is used for cascading two or more demultiplexers.
11. Demultiplexer circuits are used in clock demuliplexing, serial-to-parallel converter.

12. A combinational circuit which can express digital data given in one form to other is called code converter.
13. Code converters are classified as Encoder and Decoder.
14. Encoder is a circuit that converts a number from decimal into another number system or code.
15. Decoder is a circuit that converts a number system or code back into the decimal system.
16. A Decoder is a special case of demultiplexer. The only difference in decoder and demultiplexer is that it does not have data input.
17. Decimal to BCD encoders are widely used for interfacing external switches and logic circuit. Hexadecimal code is commonly used for entering computer data; a hexadecimal to binary converter actually interfaces the input data to computer.
18. The priority encoder includes the necessary logic which is needed to ensure that when two or more inputs are activated, only the code which corresponds to highest number input is generated at the output.
19. BCD stands for binary coded decimal; BCD code expresses each digit in decimal number by its four bit binary equivalent. BCD to Seven segment decoder is used to provide input to seven segments 'a' to 'g' of LED displays.
20. A keyboard encoder is needed to generate a code corresponding to key pressed, provides corresponding input to computer.

Exercise

[A] Multiple Choice Question :

1. A basic multiplexer principle can be demonstrated through the use of a ......
   - (a) single-pole relay
   - (b) linear stepper
   - (c) DPDT switch
   - (d) rotary switch

2. Most demultiplexers facilitate which of the following ?
   - (a) decimal to hexadecimal
   - (b) ac to dc
   - (c) single input, multiple outputs
   - (d) odd parity to even parity

3. One application of a digital multiplexer is to facilitate ......
   - (a) data generation
   - (b) parallel-to-serial data conversion
   - (c) parity checking
   - (d) code conversion

4. A multiplexed display ......
   (a) accepts data inputs from multiple lines and passes this data to multiple output lines.
   (b) accepts data inputs from one line and passes this data to multiple output lines.
   (c) uses one display to present two or more pieces of information.
   (d) accepts data inputs from several lines and multiplexes this input data to four BCD lines.

5. The following circuit is most likely a ......
   (a) comparator
   (b) multiplexer
   (c) demultiplexer
   (d) parity generator

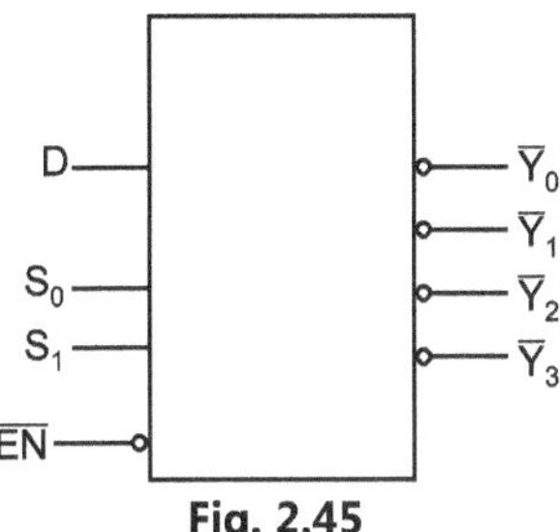

Fig. 2.45

6. Which of the figures shown below represents the exclusive-NOR gate?

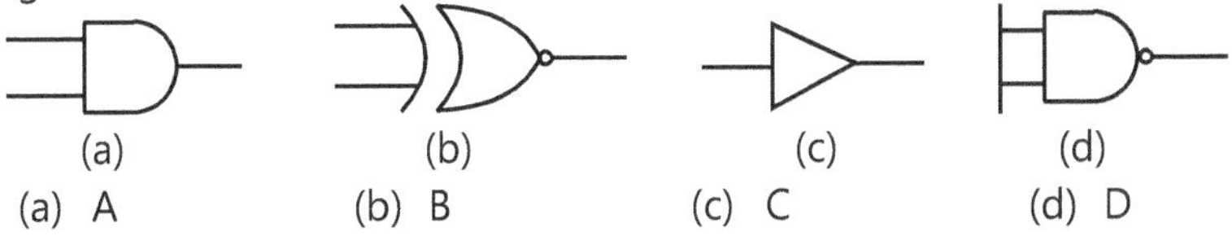

(a)	(b)	(c)	(d)
(a) A	(b) B	(c) C	(d) D

7. Which of the following statements accurately represents the two BEST methods of logic circuit simplification ?
   (a) Boolean algebra and actual circuit trial and error evaluation.
   (b) Actual circuit trial and error evaluation and waveform analysis.
   (c) Karnaugh mapping and circuit waveform analysis.
   (d) Boolean algebra and Karnaugh mapping.

8. Consider, you are confronted with a TTL circuit board containing number of IC chips. You have taken several readings at numerous IC chips, but the readings are inconclusive because of their erratic nature. Of the possible faults listed, select the one that most probably is causing the problem.
   (a) A defective output IC chip that has an internal open to V_{cc}.
   (b) An open input on the first IC chip on the board.
   (c) A solar bridge between the inputs on the first IC chip on the board.
   (d) A defective IC chip that is drawing excessive current from the power supply.

9. Which digital system translates coded characters into a more intelligible form ?
   (a) decoder
   (b) encoder
   (c) counter
   (d) display

10. A circuit that responds to a specific set of signals to produce a related digital signal output is called ......
    (a) BCD matrix
    (b) encoder
    (c) display driver
    (d) decoder

11. How many 3-line-to-8-line decoders are required for a 1-of-32 decoder ?
    (a) 1
    (b) 2
    (c) 4
    (d) 8

12. A decoder can be used as a demultiplexer by ......
    (a) connecting all data-select lines HIGH
    (b) connecting all data-select lines LOW
    (c) connecting all enable pins LOW
    (d) using the input lines for data selection and an enable line for data input

13. How many inputs are required for a 1-of-10 BCD decoder ?
    (a) 4
    (b) 8
    (c) 10
    (d) 1

14. Which type of decoder will select one of sixteen outputs, depending on the 4-bit binary input value ?
    (a) dual octal outputs
    (b) hexadecimal
    (c) binary-to-hexadecimal
    (d) hexadecimal-to-binary

15. Which of the following combinations of logic gates can decode binary 1101 ?
    (a) One 4-input AND gate
    (b) One 4-input AND gate, one OR gate
    (c) One 4-input AND gate, one inverter
    (d) One 4-input NAND gate, one inverter

16. A certain BCD-to-decimal decoder has active-HIGH inputs and active-LOW outputs. Which output goes LOW when the inputs are 1001 ?
    (a) 4
    (b) 8
    (c) 9
    (d) None. All outputs are HIGH.

17. In a BCD-to-seven-segment converter, why must a code converter be utilized?
    (a) To convert the 4-bit BCD into 7-bit code.
    (b) No conversion is necessary.
    (c) To convert the 4-bit BCD into gray code.
    (d) To convert the 4-bit BCD into 10-bit code.

18. How many outputs would two 8-line-to-3-line encoders, expanded to a 16-line-to-4-line encoder, have ?
    (a) 3   (b) 4
    (c) 5   (d) 6

19. What is the indication of a short on the input of a load gate ?
    (a) Only the output of the defective gate is affected.
    (b) There is a signal loss to all gates on the node, and the affected node will be stuck in the LOW state.
    (c) There is a signal loss to all gates on the node.
    (d) The affected node will be stuck in the LOW state.

20. What type of logic circuit is represented by the figure shown below ?

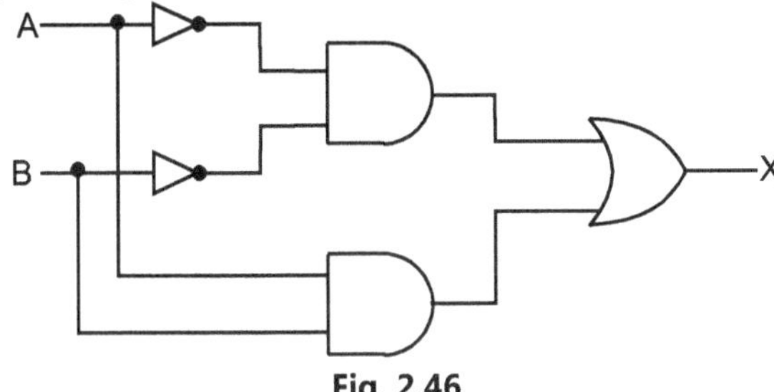

Fig. 2.46

    (a) XOR   (b) XNOR
    (c) XAND   (d) XNAND

21. Which gate is best used as a basic comparator ?
    (a) AND   (b) NAND
    (c) OR   (d) Exclusive-OR

22. If D_0, D_1, D_2, D_3 are the inputs and S_0, S_1 are select lines, then the output expression for 4 to 1 multiplexer is ......

    (a) $Y = D_0 \cdot \bar{S_1} \cdot \bar{S_0} + D_1 \cdot \bar{S_1} \cdot \bar{S_0} + D_2 \cdot S_1 \cdot \bar{S_0} + D_3 \cdot S_1 \cdot S_0$

    (b) $Y = D_0 \cdot \bar{S_1} \cdot \bar{S_0} + D_1 \cdot \bar{S_1} \cdot S_0 + D_2 \cdot S_1 \cdot \bar{S_0} + D_3 \cdot S_1 \cdot S_0$

    (c) $Y = D_0 \cdot \bar{S_1} \cdot \bar{S_0} + D_1 \cdot \bar{S_1} \cdot S_0 + D_2 \cdot \bar{S_1} \cdot \bar{S_0} + D_3 \cdot S_1 \cdot S_0$

    (d) $Y = D_0 \cdot \bar{S_1} \cdot \bar{S_0} + D_1 \cdot \bar{S_1} \cdot S_0 + D_2 \cdot S_1 \cdot \bar{S_0} + D_3 \cdot \bar{S_1} \cdot S_0$

23. The multiplexer has ......
    (a) one input, n outputs and m select lines.
    (b) n inputs, one output and m select lines.
    (c) one input, m outputs and n select lines.
    (d) m inputs, one select line and n outputs.
24. The demultiplexer has ......
    (a) one input, n outputs and m select lines.
    (b) n inputs, 1 output and m select lines.
    (c) one input, m outputs and n select lines.
    (d) m inputs, one select line and n outputs.
25. In a decoder ......
    (a) all m inputs and 1 of the n output are active at a time.
    (b) 1 of the n inputs and all m outputs are active at a time.
    (c) 1 of the n inputs and 1 of the n outputs are active at a time.
    (d) all inputs and outputs are active at a time.
26. Decoders can be used as ......
    (a) multiplexer   (b) demultiplexer
    (c) encoder   (d) display
27. Scan time for matrix keyboard encoder is ...... than scan time for linear encoder.
    (a) faster   (b) slower
    (c) same   (d) none of above
28. The 7-segment LED display can be ......
    (a) common anode type   (b) common cathode type
    (c) both (a) and (b)   (d) none of (a) and (b)
29. One application of a digital multiplexer is to facilitate ......
    (a) data selector   (b) serial-to-parallel conversion
    (c) data generation   (d) parity checking
30. The BCD number for decimal 18 is ......
    (a) 00011000   (b) 10101010
    (c) 11111000   (d) 00010000

Answers

1. (d)	2. (c)	3. (b)	4. (c)	5. (c)	6. (b)	7. (d)	8. (b)
9. (a)	10. (b)	11. (c)	12. (d)	13. (a)	14. (b)	15. (c)	16. (c)
17. (a)	18. (b)	19. (b)	20. (b)	21. (d)	22. (b)	23. (b)	24. (a)
25. (a)	26. (b)	27. (a)	28. (c)	29. (a)	30. (a)		

[B] True or False :

1. Combinational circuits requires clock for their operation.
2. In case of combinational circuit output depends upon past input and output.
3. Half adder adds bits at a time.
4. IC 7483 is used for addition of two 4-bit numbers.
5. Magnitude compressors are used to compare binary numbers.
6. Multiplexers have one input and many outputs.
7. Demultiplexers can be used as decoders.
8. EX-OR gates are used for parity checking.
9. Encoders are used to convert more general code into desired code.
10. Multiplexer is a logic circuit that selects one out of several data inputs to a single output.

Answers						
1. False	2. False	3. True	4. True	5. True	6. False	7. True
8. True	9. True	10. True				

[C] Short Answer Questions :

1. Explain the working of following multiplexers with logic diagram and truth table :

   (a) 2 to 1 multiplexer.

   (b) 4 to 1 multiplexer.

2. Design a 4 : 1 multiplexer using only NAND gates.

3. Construct 4 : 1 multiplexer using 2 to 1 multiplexer, give its function table.

4. Explain the working of following demultiplexers with logic diagram and truth table :

   (a) 1 to 2 demultiplexer.

   (b) 1 to 4 demultiplexer.

5. With the help of logic diagram and truth table, describe the action of decimal to BCD encoder.

6. Write a note on priority encoder.

7. Describe the design procedure and hence draw the logic diagram of hex to binary encoder.

8. With the help of neat diagram, explain the action of keyboard encoder.

9. Describe the working of BCD to decimal decoder with neat diagram.

10. What do you mean by seven segment display ? Which are the different types of seven segment display ?

11. Describe the BCD to seven segment display decoder/driver.

12. Draw the circuit diagram of decoder driving common anode type of display.

[D] Long Answer Questions :

1. Define the terms multiplexers and demultiplexers.
2. Describe the functions of strobe and select inputs of a multiplexer.
3. Explain the difference between a decoder and demultiplexer.
4. State the applications of multiplexers and demultiplexers.
5. Define the terms encoders and decoders.

6. Explain the functions of LT, BI/RBO and $\overline{\text{RBI}}$ inputs of seven segment decoder IC.

❑❑❑

Chapter **3**...

Sequential Logic Circuits

The first ever recorded **currency-counting machine** was **invented** in 1958 in China, **created** by a small-time banker Zhi Tian Sie.

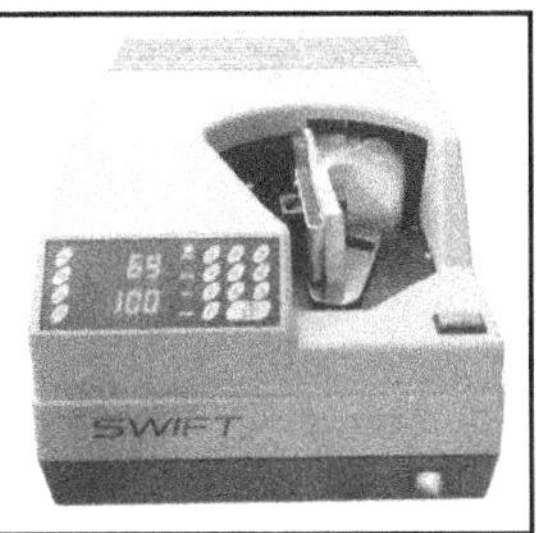

Godrej Swift Turbo Desktop Bundle Note Counting Machines

3.1 Introduction to Flip-Flops

- The logic gates and arithmetic circuits (chapter 2) have their outputs totally depend on the inputs.

- In these circuits, if the input changes, the output also changes. However, in a digital system, we require the circuits whose output once set should remain unchanged even if the input is changed.

- Such a device can be used to store information in the form of binary number.

- The cross-coupled gates used to store binary data are known as *flip-flops*. There are different types of flip-flops.

- Flip-flops are basic building blocks of counters, shift registers and memory devices.

- Digital electronics is classified into combinational logic and sequential logic.

- In the previous unit, we have learned the concepts of combinational logic circuits where, output depends on the current inputs.

- Sequential logic output depends on current as well as past inputs, so we need memory elements such as flip-flops and latches.

3.2 The S-R Flip-Flop

- Fig. 3.1 shows the circuit diagram of SR flip-flop using transistor. There is cross coupling of each collector to the opposite base. This cross coupling results in positive feedback.

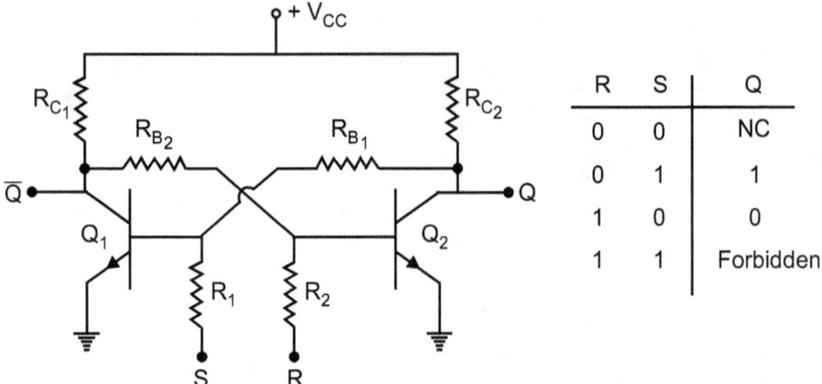

R	S	Q
0	0	NC
0	1	1
1	0	0
1	1	Forbidden

(a) S-R flip-flop using transistor   **(b) Truth table**

Fig. 3.1

- As shown in Fig. 3.1 (a), the circuit has two inputs S and R known as set and reset respectively and outputs Q and $\bar{Q}$. The resistors are such that $R_1 = R_2$, $R_{C_1} = R_{C_2}$ and $R_{B_1} = R_{B_2}$.

Case (i) S = 1, R = 0 :

- By the application of high voltage to S input, since both Q_1 and Q_2 are NPN transistors, transistor Q_1 goes into saturation, its collector voltage becomes low, hence base of Q_2 will be low which makes Q_2 cut-off and thus collector of Q_2 remains at V_{CC} or $Q = 1$.
- Since collector of Q_2 is connected to base of Q_1, it will keep Q_1 into saturation i.e. state $Q = 1$, $\bar{Q} = 0$ is permanently stored as long as S = 1, R = 0 and hence such system is also known as *latch*.

Case (ii) S = 0, R = 1 :

- Due to R = 1, transistors being NPN, Q_2 goes into saturation, which makes collector of Q_2 low which drives Q_1 to cut-off making Q = 0 and $\bar{Q} = 1$.

Case (iii) S = 0, R = 0 :

- This means no trigger is applied. In this case, due to circuit imbalance, either Q_1 becomes ON or Q_2 becomes ON and other remains OFF.

- Even though collector resistors are same in practice, they are slightly different e.g. if R_{C_1} is slightly smaller than R_{C_2}, Q_1 conducts more than Q_2, collector voltage of Q_1 decreases which makes Q_2 more OFF increasing its collector voltage which further increases base voltage of Q_1.

- This is cumulative action making Q_1 ON and Q_2 OFF. Whereas when R_{C_2} is slightly smaller than R_{C_1}, Q_2 is made ON. Thus, when S = 0, R = 0 either of the two transistors becomes ON and there is no change in the state of the flip-flop.

Case (iv) S = 1, R = 1 :

- When trigger is applied to both inputs simultaneously then both the transistors try to become ON and try to go into saturation which

  make Q and $\bar{Q}$ low but Q and $\bar{Q}$ must be opposite and hence user should not apply S = 1, R = 1 which is not allowed or forbidden condition.

- Thus, in S-R flip-flop, when S and R are different, output follows S input. The S-R flip-flop can also be build using universal gates.

3.2.1 S-R Flip-Flop using NAND Gates

- Flip-flop can be obtained using any type of gates, but since NAND and NOR are the universal gates, we will see flip-flop using these gates. Fig. 3.2 shows S-R flip-flop using NAND gates.

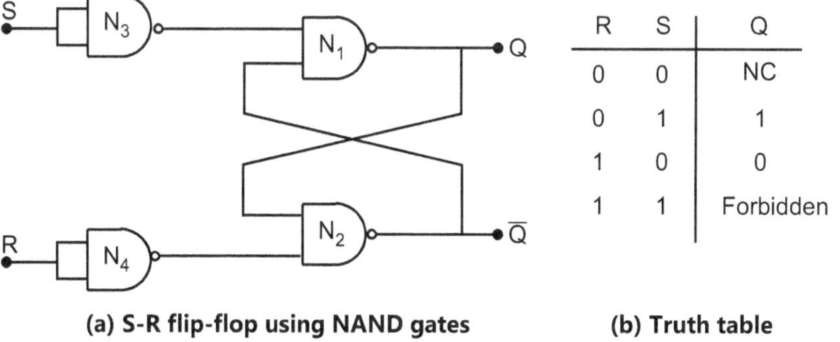

R	S	Q
0	0	NC
0	1	1
1	0	0
1	1	Forbidden

(a) S-R flip-flop using NAND gates　　　　　**(b) Truth table**

Fig. 3.2

- The circuit uses four NAND gates. N_1 and N_2 are two inputs whereas N_3 and N_4 are used as an inverter.

- It has two inputs S and R and two outputs Q and $\bar{Q}$. The outputs are cross coupled to the input; that is known as *feedback*.

- This cross coupling or feedback is responsible for memory of the circuit and that is the general feature of all flip-flop circuits. Since there are two inputs, there are four possible cases.

Case (i) S = 0, R = 1 :

- Then output of N_3 is 1 and output of N_4 is 0. Whatever may be the initial output for N_2, one input is zero, so $\bar{Q}$ = 1, for N_1 both inputs are 1, so Q becomes 0.

Case (ii) S = 1, R = 0 :

- Then output of N_3 is 0 and output of N_4 is 1. Whatever may be initial output for N_1, one input is 0, so output Q becomes 1 which is fed back to N_2, for N_2 both inputs are 1 which make $\bar{Q}$ = 0.

Case (iii) S = 0, R = 0 :

- Outputs of both N_3 and N_4 will become 1. If previously Q = 0 for N_2 one input is 0 which makes $\bar{Q}$ = 1, that 1 is fed back to N_1, for N_1 both inputs are 1 which keep the output 0 which is assumed.
- If we assume previous output 1, it will remain the same that means when S = 0, R = 0 there is no change in the state of the flip-flop.

Case (iv) S = 1, R = 1 :

- For this situation, outputs of N_3 and N_4 will become 0, one input of both NAND gates N_1 and N_2 becomes 0 which tries to make Q and $\bar{Q}$ both 1 but that is contradictory to our definition. So S = 1, R = 1 is the forbidden condition.
- Note that N_3 and N_4 are added only for the convenience. They may be absent. In the   Fig. 3.2 (a) they just invert the input.
- However, if they are absent, the output Q must be named to the gate whose one input is R. Fig. 3.3 shows S-R flip-flop using two NAND gates along with the truth table.

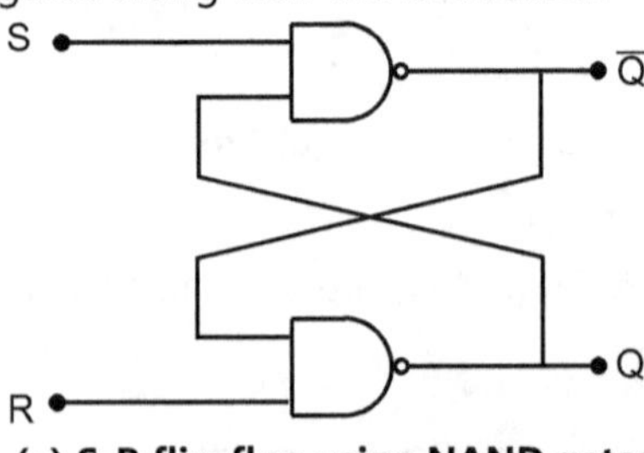

R	S	Q
0	0	Forbidden
0	1	1
1	0	0
1	1	NC

(a) S-R flip-flop using NAND gates   **(b) Truth table**

Fig. 3.3

- The S-R flip-flop can also be build using NOR gates. The S-R flip-flop using NOR gates is shown in Fig. 3.4 along with the truth table.

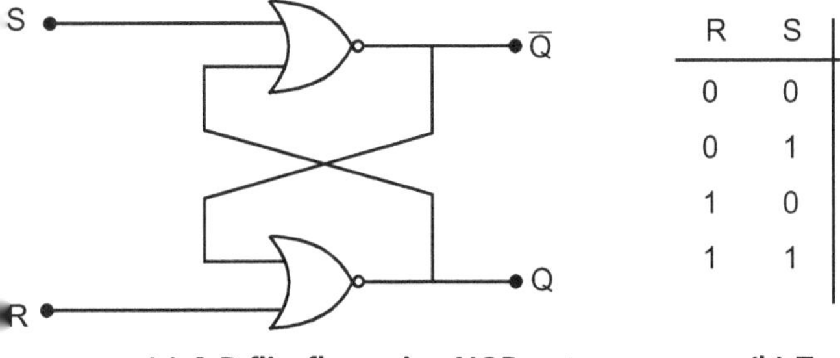

R	S	Q
0	0	NC
0	1	1
1	0	0
1	1	Forbidden

(a) S-R flip-flop using NOR gates   **(b) Truth table**

Fig. 3.4

- We can see from the truth table that, when S and R are different, output follows the S input in both the cases. However, for NAND gate S-R flip-flop 00 is forbidden condition, whereas for NOR gate S-R flip-flop, 11 is the forbidden condition.

3.3 The Clocked S-R Flip-Flop

- In a sequential system, it is required to set or reset the flip-flop in synchronism with clock pulses. This is achieved in clocked S-R flip-flop.

- The S-R flip-flop is modified to clocked S-R flip-flop by changing N_3 and N_4 in Fig. 3.2 (a) to two input NAND gates and by applying the clock pulses simultaneously to N_3 and N_4. Fig. 3.5 shows the clocked S-R flip-flop.

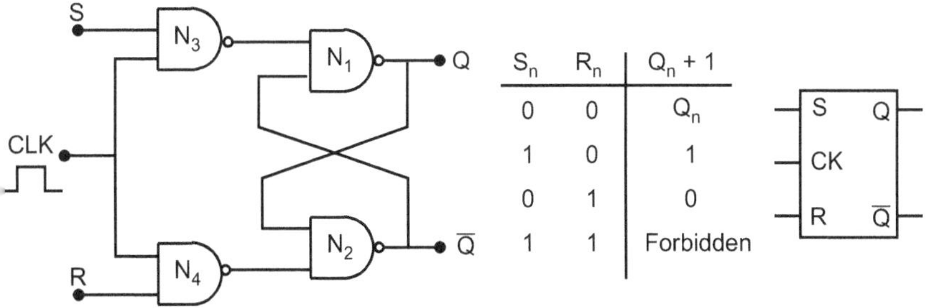

S_n	R_n	$Q_n + 1$
0	0	Q_n
1	0	1
0	1	0
1	1	Forbidden

(a) Clocked S-R flip-flop   **(b) Truth table**   **(c) Logic symbol**

Fig. 3.5

- If the clock pulse is absent, CK = 0. Whatever may be S and R, outputs of N_3 and N_4 are 1. If Q = 1 it remains 1 whereas if Q = 0 it remains 0. The flip-flop does not change the state. The circuit is equivalent to a latch.

- If the clock pulse is present, CK = 1. If S = 0, R = 0 then outputs of N_3 and N_4 are 1 and the output of the flip-flop does not change. Hence, after the pulse, output remains same as before the pulse.
- If we denote the inputs S_n and R_n before the pulse then output after the pulse is $Q_{n+1} = Q_n$ and this is indicated in the first row of the truth table.
- If CK = 1, S_n = 1 and R_n = 0 then output of N_3 is 0, output of N_4 is 1, which make Q = 1. Hence, after the clock pulse, we find Q_{n+1} = 1.
- If CK = 1, S_n = 0 and R_n = 1 then we find Q_{n+1} = 0. This confirms second and third rows of the truth table.
- If CK = 1, S_n = 1 and R_n = 1 then outputs of NAND gates N_3 and N_4 are both 0. Hence whatever may be previous output one input of N_1 and N_2 is 0, so that outputs of both N_1 and N_2 must be 1, but according to our definition, Q and $\bar{Q}$ must be complement.
- In practice, whether N_1 input or N_2 input rises faster from 0 to 1 and depending upon circuit parameter asymmetries either the state Q = 0 or Q = 1 is obtained.
- This state is said to be indeterminate, ambiguous or undefined and the condition S_n = 1, R_n = 1 is forbidden and should not be applied. This ambiguity present in the truth table is overcome by J-K flip-flop.

3.4 The J-K Flip-Flop

- J-K flip-flop can be obtained from S-R flip-flop along with two AND gates G_1 and G_2. Data input J and the output $\bar{Q}$ are applied to G_1.
  Since its output generates S, then S = $J\bar{Q}$.
- Similarly, data input K and the output Q are applied to G_2 and hence R = KQ. Fig. 3.6 shows S-R flip-flop converted into J-K and its truth table.

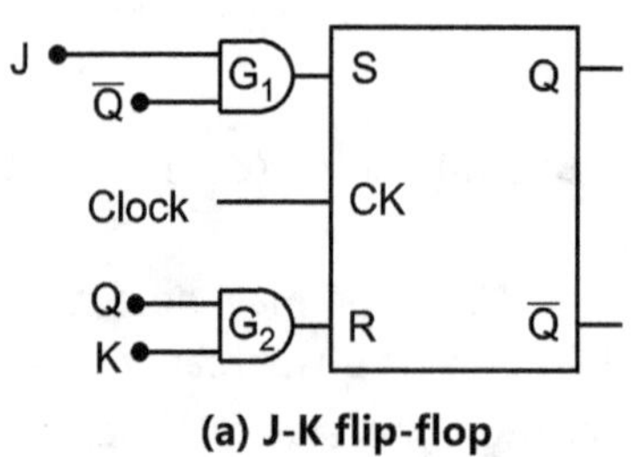

<table>
<tr><td></td><td>(a) J-K flip-flop</td><td></td></tr>
</table>

J_n	K_n	$Q_n + 1$
0	0	Q_n
1	0	1
0	1	0
1	1	$\bar{Q}_n$

(b) Truth table

Fig. 3.6

- For two J and K inputs there are four possibilities. At the same time there are two possible states for Q which generate eight possibilities.

- The following Table 3.1 has eight rows from J_n, K_n, Q_n and $\bar{Q}_n$, the S_n and R_n are calculated and entered into 5th and 6th columns of the table. Using these values of S_n and R_n and referring to the S-R flip-flop, 7th column is obtained and finally column 8 follows from column 7.

Table 3.1 : Truth table of Fig. 3.6

Row / Column	1 J_n	2 K_n	3 Q_n	4 $\bar{Q}_n$	5 S_n	6 R_n	7 Q_{n+1}	8
1	0	0	0	1	0	0	Q_n	Q_n
2	0	0	1	0	0	0	Q_n	
3	1	0	0	1	1	0	1	1
4	1	0	1	0	0	0	Q_n	
5	0	1	0	1	0	0	Q_n	0
6	0	1	1	0	0	1	0	
7	1	1	0	1	1	0	1	$\bar{Q}_n$
8	1	1	1	0	0	1	0	

- Thus, we see that the first three rows of J-K flip-flop truth table are same as that of S-R flip-flop truth table. When the data inputs J and K are high, the output gets complemented by the clock pulse.
- In case of S-R flip-flop, we cannot apply S = 1, R = 1 but in case of J-K flip-flop, we can apply J = 1, K = 1.

3.4.1 J-K Flip-Flop With Only NAND Gates

- Fig. 3.6 (a) shows J-K flip-flop which uses AND gates. However, we can also obtain J-K flip-flop only using NAND gates.
- Fig. 3.7 shows J-K flip-flop using only NAND gates. N_1 and N_2 are two input NAND gates whereas N_3 and N_4 are three input NAND gates.
- If J = 0, K = 0, outputs of N_3 and N_4 are 1, since one input of N_3 and

N_4 is 0. If Q = 0 and $\bar{Q}$ = 1 for N_2 one input is 0, so its output remains 1 and for N_1 both inputs are 1 so gives same output 0 as assumed. Thus, output remains in the previous state.

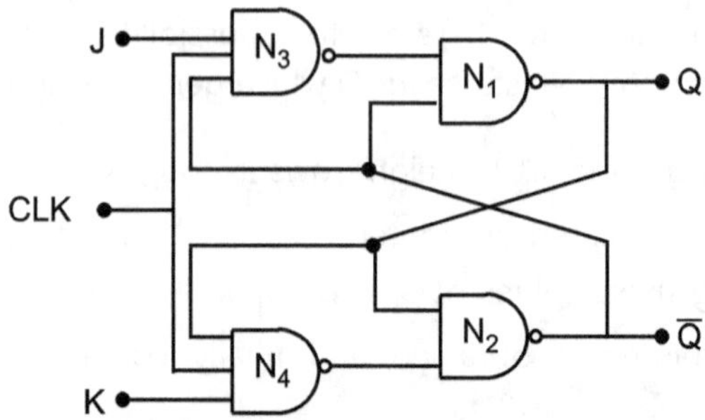

Fig. 3.7 : J-K flip-flop using only NAND gates

- If J = 0, K = 1 and CLK = 1, for N_3 one input is 0, so output is 1. For N_4, if Q = 1 all inputs are 1, so output is 0 then for N_2 one input 0, so $\overline{Q}$ becomes 1. For N_1 both inputs are 1, so output Q becomes 0. So output becomes same as J input even if J = 1 and K = 0.

- If J = 1, K = 1 and CLK = 1. If Q = 1 for N_4 all inputs are 1, so output of N_4 is 0. For N_3 one input is 0, so output is 1. For N_2 one input is 0, so $\overline{Q}$ becomes 1. For N_1 both inputs are 1, so output becomes 0 i.e. if we assume Q = 0 it becomes 1. If we assume Q = 1, it becomes 0. Thus, when J = 1, K = 1 output becomes complement of previous state.

- Thus, for all possible values of J and K, the circuit follows the truth table of J-K flip-flop.

3.4.2 Preset and Clear

- In case of J-K flip-flop when J = 1, K = 1 and clock is applied, output becomes complement of previous state. But the previous state is arbitrary, it may be 0 or 1.

- It is possible to assign initial state to the flip-flop. We can clear the flip-flop or set the flip-flop by addition of two or more inputs to the flip-flop known as *preset* and *clear*.

- If NAND gates N_1 and N_2 of Fig. 3.7 are converted into three inputs as shown in Fig. 3.8, we can assign initial state to the flip-flop.

- If we want to clear the flip-flop, apply Pr = 1, Cr = 0 and CLK = 0. Since Cr is 0 the output of N_2 is $\overline{Q}$ = 1. Since CLK = 0, the output of N_3 is 1 and hence all inputs to N_1 are 1 which give Q = 0 as required.

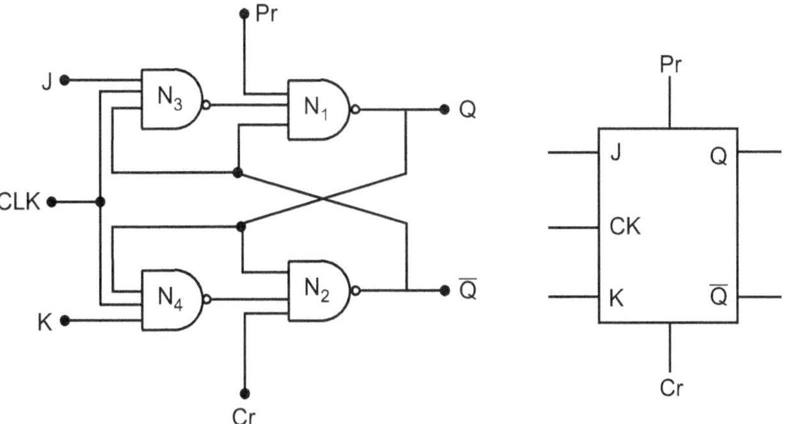

(a) J-K flip-flop with preset and clear   **(b) The logic symbol**

Fig. 3.8

- Similarly, if we want to preset the latch or flip-flop, apply Pr = 0, Cr = 1, CLK = 0. The preset and clear input are called *direct* or *asynchronous input*, since they are not in synchronous with the clock and may be applied before the clock or in between the clock pulses.

- Once the state to the flip-flop is assigned, Pr = 1, Cr = 1 must be maintained, that is for the normal operation of the flip-flop, Pr = 0 and Cr = 0 must not be used since it leads to ambiguous state at the output.

Table 3.2 summarises the operation of preset and clear input.

Table 3.2 : Preset and clear requirement

CLK	Cr	Pr	Remark
0	0	1	To clear the flip-flop
0	1	0	To set the flip-flop
1	1	1	For normal operation
×	0	0	Not allowed

3.4.3 The Race Around Condition

- The truth table of Fig. 3.6 (b) of J-K flip-flop assumes that the inputs are independent of the output.

- However, because of the feedback connection at the input there is difficulty with the J-K flip-flop constructed as shown in Fig. 3.6 or 3.7. Consider for example, J = 1, K = 1 and Q = 0.

- When the clock pulse is applied, the output becomes Q = 1. This change takes place after a time interval Δt equal to the propagation delay through NAND gates in series.

- If clock pulse is still there since J = 1, K = 1 and Q = 1, the output will become 0. Thus, for the duration t_p of the clock pulse, the output will oscillate back and forth between 0 and 1, and at the end of the pulse, the value of Q may be 0 or 1 i.e. ambiguous. This situation is known as race around condition.

- It can be avoided if $t_p < \Delta t$. However, in practice, Δt is very small. There are two possible solutions: one is to convert wide duration clock pulse into spikes using differentiator circuit and then apply to clock input or construct another circuit known as master slave J-K flip-flop.

3.4.4 Concept of Triggering Levels

- In a simple clocked flip-flop when the input of a flip-flop is changed, output will change as long as clock is high, whether the flip-flop is S-R or J-K.

- When the clock goes low, the state present at the input will be stored or latched, in case of S-R flip-flop, whereas in case of J-K flip-flop output gets complemented.

- Usually, it is necessary to store the input of flip-flop instantaneously and for that we need a very narrow clock pulse. Such narrow clock pulse is also required in case of J-K flip-flop to avoid race around condition.

- The square pulse can be converted into positive and negative going spikes using R-C differentiator circuit. Fig. 3.9 shows such circuit.

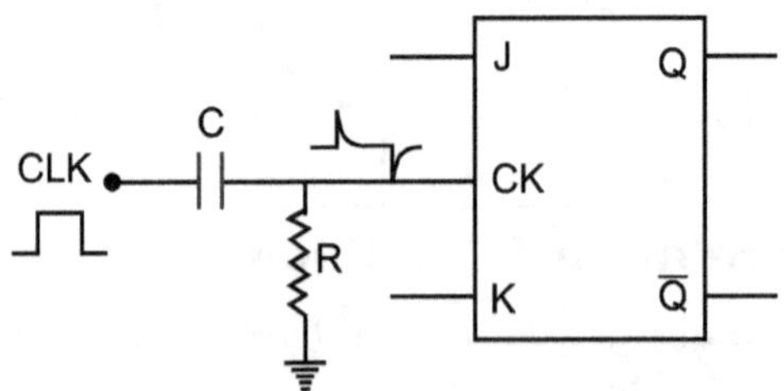

Fig. 3.9 : J-K flip-flop with edge triggering

- The values of R and C are so chosen that RC time constant is much smaller than the clock pulse width and hence such circuit converts pulses into positive and negative going spikes.

- The narrow positive spikes enable the internal gates for an instant. The narrow negative spikes does nothing. Thus, during positive spikes gate is enabled, flip-flop follows its truth table. This kind of operation is called *edge triggering* because flip-flop responds only when the clock is in transition.

- In the Fig. 3.9, the triggering occurs on the positive going edge of the clock. So it is referred as positive edge triggering.

- Sometimes triggering on negative edge is better suited or more suitable to the application. In this case, an internal inverter can complement the clock pulse before it reaches to the NAND gate.

- The trailing/falling edge of the clock activates the gates, this is known as *negative edge triggering*. Fig. 3.10 shows the circuit symbol of positive and negative edge triggered flip-flops.

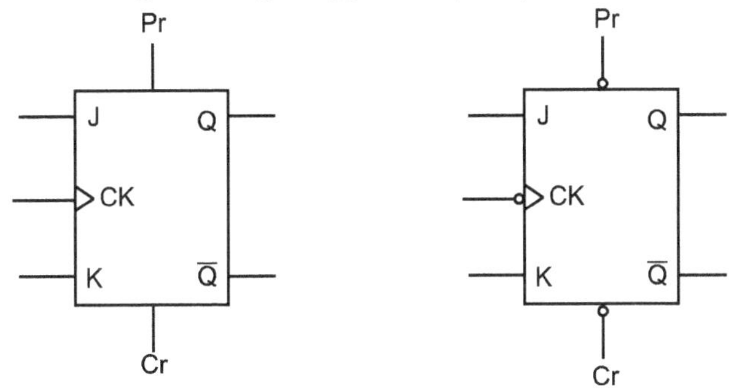

(a) Positive edge triggering   **(b) Negative edge triggering**

Fig. 3.10 : Edge triggering

3.5 Master Slave J-K Flip-Flop

- Fig. 3.11 shows the two S-R flip-flops cascaded with feedback from the output of the second to the input of the first.

- The first stage is known as *master* whereas the second as *slave*. Clock pulses are applied to the master, the same are inverted and then applied to the slave.

- As seen previously for normal operation, Pr = 1, Cr = 1 must be held. For Pr = 1, Cr = 1 and CK = 1, the master is enabled and its operation follows the J-K truth table.

- At the same time, since $\overline{CK}$ = 0, the slave remains disabled or cannot change the state. So output Q_n does not change the state during t_p

and hence the race around difficulty is removed, after the pulse passes CK = 0, so that the master is inhibited and $\overline{CK}$ = 1, which causes slave to be enabled.

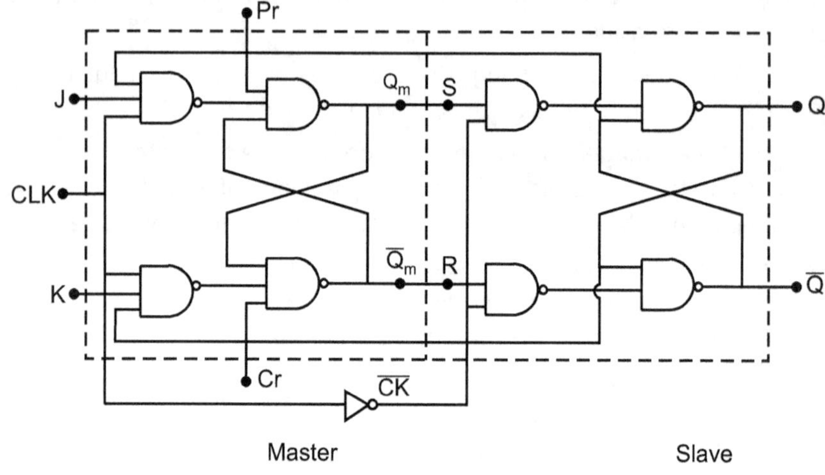

Fig. 3.11 : Master slave J-K flip-flop

- The slave is on SR flip-flop and follows its truth table. If S = 1 and R = 0 then Q = 1 or if S = 0 and R = 1 then Q = 0.

- Thus, in case of M-S J-K flip-flop, during a clock pulse output Q does not change but Q_m follows J-K logic and at the end of the pulse, the value of Q_m is transferred to Q.

3.6 D Flip-Flop

- J-K flip-flop is modified into D (delay) flip-flop by adding an inverter between J and K input, so that K is always the complement of J. The input is named as D input. Fig. 3.12 shows logic circuit, logic symbol and the truth table of D type flip-flop.

- If D_n is 1 then J_n = 1, K_n = 0 and hence Q_{n+1} = 1.

- If D_n is 0 then J_n = 0, K_n = 1 and then Q_{n+1} = 0.

- The output Q_{n+1} after the pulse equals the input D_n before the pulse and therefore D flip-flop is used to provide delay. The bit on the D line is transferred to the output at the next clock pulse and hence flip-flop functions as 1-bit delay device.

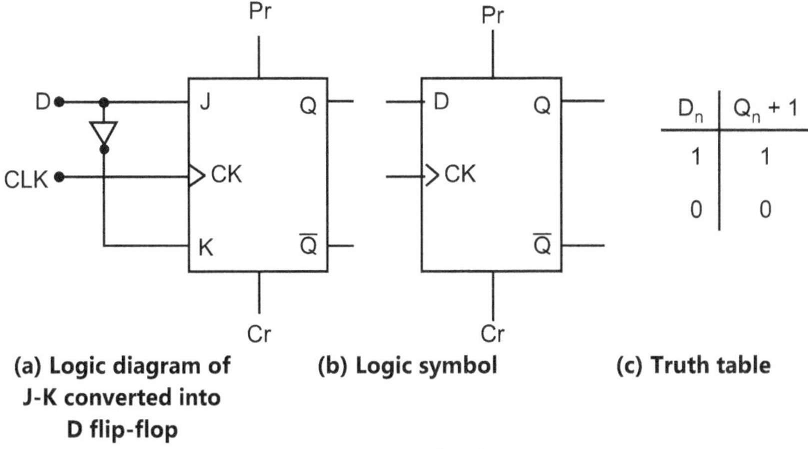

(a) Logic diagram of J-K converted into D flip-flop   **(b) Logic symbol**   **(c) Truth table**

Fig. 3.12 : D flip-flop

- The S-R flip-flop can also be converted into D type by adding an inverter between S and R and then there is no ambiguous state. Since S = R is not possible.

3.7 T Flip-Flop

- The J-K flip-flop is modified into T by connecting J and K together. This unit changes the state with each clock pulse, acts as a toggle switch and hence the name. Fig 3 .13 shows the T-flip-flop.

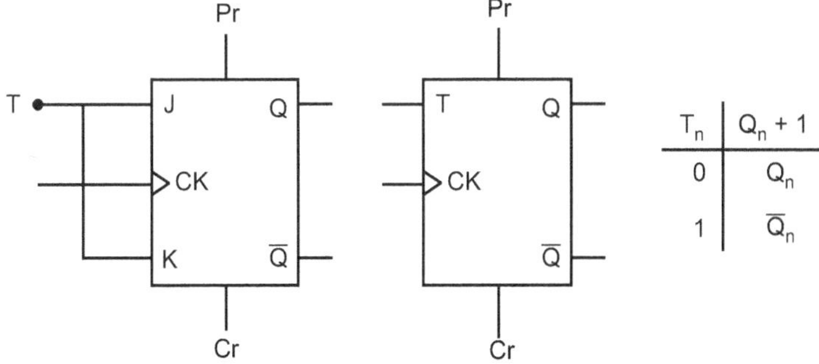

(a) J-K converted into T flip-flop   **(b) Logic diagram**   **(c) Truth table**

Fig. 3.13 : T flip-flop

- When J = K = 1 then $Q_{n+1} = \bar{Q}_n$ and when J = K = 0, then $Q_{n+1} = Q_n$. According to the truth table of J-K flip-flop, T flip-flop is also known as *complementing flip-flop*.
- If T = 1 and the clock pulses as shown in Fig. 3.14 are applied, the flip-flop behaves as divide by 2 circuit because it gives 0 to 1 transition at first clock pulse and 1 to 0 at the next clock pulse i.e. two input pulses are required to have one output pulse.

- If the clock input square wave has frequency of 10 kHz, the output at T flip-flop will have frequency 5 kHz. Thus, one flip-flop does the operation of frequency division by two.

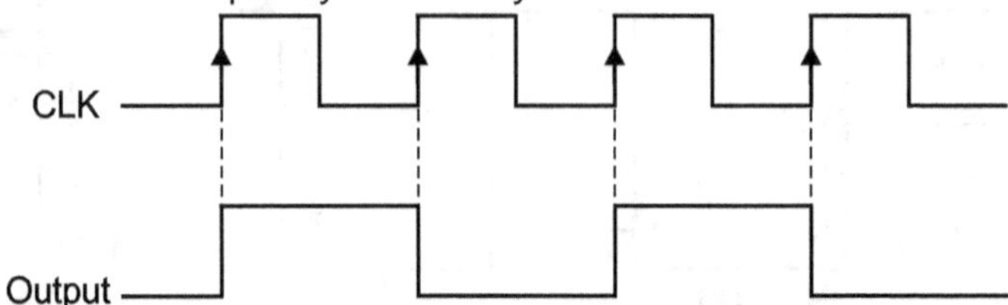

Fig. 3.14 : T flip-flop output

- The flip-flop forms the basic building blocks of one of the important digital systems that is counter which is studied in the next section.

3.8 Introduction to Counters

- One of the most versatile subsystems in digital electronics is the Counter. *Counter* is basically constructed using flip-flops and is an example of *sequential circuits*. Counters count the number of clock pulses arriving at its input.
- The output of the counter is binary or the BCD equivalent of the number of clock pulses applied at its input. Depending upon the nature of the output, one way to classify the counters is the binary counters and the BCD counters.
- Counters are basically useful to count the clock pulses applied at the input. Since clock pulses occur at known intervals, therefore counter can be used as an instrument for measuring time and therefore period or frequency.
- Many physical parameters or events can be converted into pulses and hence counter can be used to indicate temperature, vibrations etc. or to count events.
- There are two types of counters: Asynchronous or Ripple counter and Synchronous or Parallel counters.

3.9 Flip-Flop as a Memory Device

- Flip-flops are used to store binary data. One flip-flop stores on bit.
- For storage of 4-bit binary number we require 4 flip-flops.
- In order to store logic '1' we have to set the flip-flop. In order to store logic '0' we have to reset the flip-flop. Flip-flop stores the bit's as long as power supply is ON.
- In digital systems like computer we need to store binary data. The digital element used to store data is known as memory. In case of computer, memory is one important element. Such memory element contains group of flip-flops arranged in such a way that we can read or write the binary data into it.

3.10 Asynchronous, Mod 16 Binary Counter

- If the output of a counter is proportional to the binary equivalent of number of clock pulses applied at its input, it is known as *binary counter*.
- Fig. 3.15 shows the circuit of 4-bit binary counter. As seen, it is obtained by cascading the flip-flops such that output of previous flip-flop is connected to the clock input of the next flip-flop.

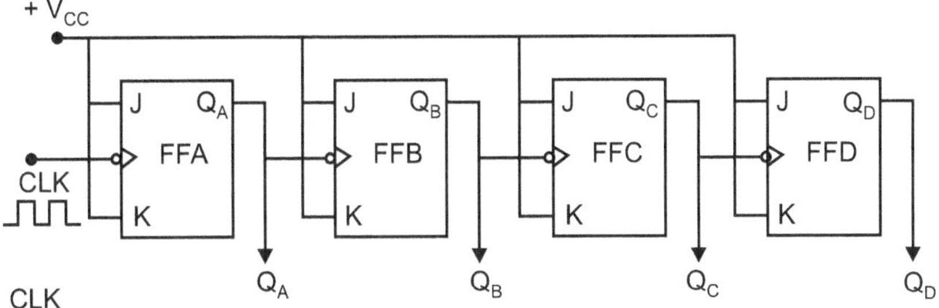

Fig. 3.15 : 4-bit binary counter

- Note that the output of A flip-flop drives the B flip-flop. B flip-flop drives the C flip-flop and C flip-flop drives the D flip-flop.
- The J and K inputs of all flip-flops are connected to V_{CC}, i.e. J-K flip-flop is converted to T type in which each flip-flop changes the state with a negative transition of its clock input.
- When clock pulse is applied and it goes negative, flip-flop A changes the state. A flip-flop has to change the state to trigger B flip-flop.
- B has to change the state before it can trigger C. C has to change the state before it can trigger D.
- The trigger moves through the flip-flops like a ripple in water. The counters in which output of one flip-flop drives the another are called as *ripple counters* or *asynchronous counters*.

Working :

- Before application of any clock pulse, initially all flip-flops are reset to produce the output DCBA = 0000.
- All flip-flops can be reset by applying clear pulse simultaneously to all flip-flops.
- When the first clock pulse is applied, flip-flop A changes its state on the negative going part of the pulse.
- Flip-flop A output goes from logic 1 to 0 which is positive change and hence flip-flop B does not responds to it. Since flip-flop B has not changed its state, flip-flop C will not change its state and since C

has not changed, D will not change. So at the end of first clock cycle, the output condition is DCBA = 0001.

- When the second clock pulse is applied, flip-flop A changes its state at the negative edge of pulse and A flip-flop output goes from 1 to 0, a negative change.
- This negative going change triggers the B flip-flop, therefore, B goes from 0 to 1. This positive going change in B has no effect on C, since C has not changed, no change in D flip-flop. So at the end of second clock pulse, the output condition of four flip-flops is DCBA = 0010.
- When the third clock pulse is applied, flip-flop A changes its state from 0 to 1; this is positive going change and has no effect on the B flip-flop. B has not changed and hence C and further D will not change. So at the end of third clock pulse, the output condition of four flip-flops is DCBA = 0011.
- Thus, we can see that the output condition of four flip-flops is binary equivalent of number of clock pulses applied at its input. Table 3.3 shows the truth table of 4-bit binary counter.

Table 3.3

Clock pulses	D	C	B	A	
0	0	0	0	0	
1	0	0	0	1	
2	0	0	1	0	
3	0	0	1	1	
4	0	1	0	0	
5	0	1	0	1	
6	0	1	1	0	
7	0	1	1	1	
8	1	0	0	0	
9	1	0	0	1	
10	1	0	1	0	
11	1	0	1	1	
12	1	1	0	0	
13	1	1	0	1	
14	1	1	1	0	
15	1	1	1	1	
16	0	0	0	0	← Reset
17	0	0	0	1	← Recounting

- Since the circuit gives the output which is binary equivalent of number of clock pulses applied at its input, it is known as *binary counter.*

- At the end of 15th clock pulse, flip-flop's output is DCBA = 1111 and the circuit counts upto maximum of 15.

- When 16th clock pulse is applied, A goes from 1 to 0, a negative going change. This causes B to go from 1 to 0, another negative going change, this forces C to go from 1 to 0. This negative going change causes D to go from 1 to 0, so the entire counter resets and the output condition of flip-flop becomes DCBA = 0000.

- When 17th clock pulse is applied, counter starts recounting. Fig. 3.16 shows the waveforms associated with 4-bit binary counter of Fig. 3.15.

- Since there are 16 different states at the output, this circuit is known as Mod 16 counter.

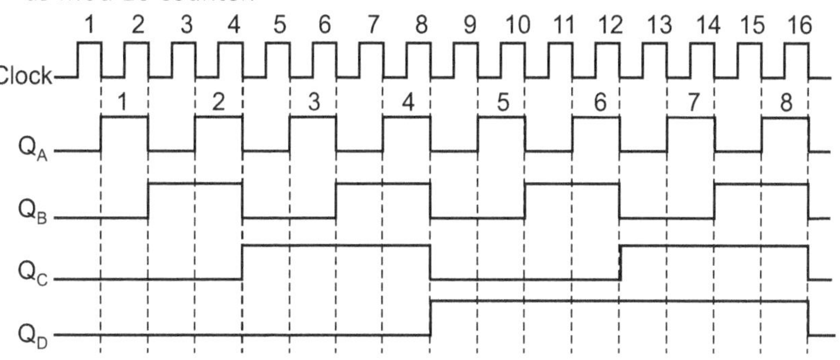

Fig. 3.16 : Waveforms of 4-bit binary counter

- From the waveforms, we see that the waveform at Q_A is one half the frequency of clock, at Q_B is one fourth of the clock frequency, at Q_C is one eighth of the clock frequency and at Q_D one sixteenth of the clock frequency. Thus, counter is also a frequency divider.

3.11 Down Counter

- The counters we have studied in section 3.10 are also known as *up counters*, since upon receiving the next clock pulse, the content of counter increases by one. The count sequence of such counters are 0, 1, 2, 3, ... n.

- It is sometimes useful to have counter which can count in downward sequence i.e. the count sequence is n, n – 1, n – 2, ...1, 0. Upon receiving the next clock pulse if the content of counter decreases then it is known as *down counter.*

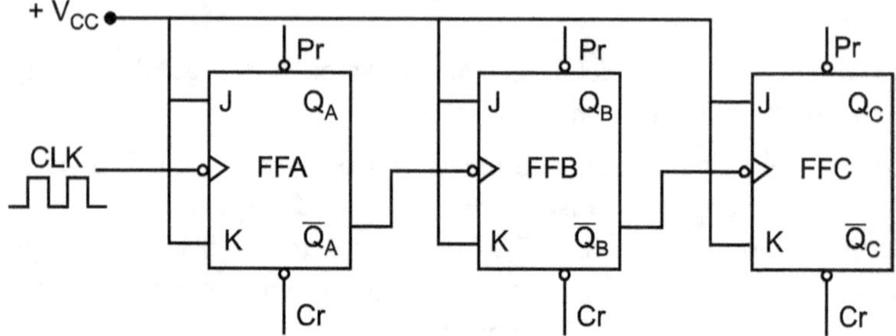

Fig. 3.17 : 3-bit down counter

- Fig. 3.17 shows the circuit of 3-bit down counter. The 3-bit down counter is obtained using 3 flip-flops, the system clock is still applied at flip-flop A. The $\overline{Q}_A$ is used to drive flip-flop B and $\overline{Q}_B$ is used to drive flip-flop C.

- Initially, all flip-flops are reset by applying clear pulse simultaneously to all flip-flops. When the 1st clock pulse is applied, flip-flop A toggles at negative edge of the pulse and Q_A becomes 1.

- As Q_A goes from 0 to 1, $\overline{Q}_A$ goes from 1 to 0 which is negative change and it triggers flip-flop B causing Q_B to go from 0 to 1, $\overline{Q}_B$ becomes 1 to 0 which is negative change and it further triggers flip-flop C.

- So at the end of 1st clock pulse, the output condition of three flip-flops is,

$$Q_C\, Q_B\, Q_A \;=\; 111$$

- When 2nd clock pulse is applied at the negative edge, flip-flop A toggles, Q_A output goes from 1 to 0, $\overline{Q}_A$ goes from 0 to 1 which is positive change and it does not affects flip-flop B. Since flip-flop B has not changed, C will not change, so at the end of 2nd clock pulse,

$$Q_C\, Q_B\, Q_A \;=\; 110$$

- When 3rd clock pulse is applied, the output condition of the flip-flop will be $Q_C\, Q_B\, Q_A = 101$ and so on i.e. after receiving next clock pulse, content of counter decreases by one i.e. the down counter. Table 3.4 gives the truth table of 3-bit down counter.

Table 3.4

Count	Q_C	Q_B	Q_A	$\overline{Q}_C$	$\overline{Q}_B$	$\overline{Q}_A$	$\overline{\text{Count}}$
7	1	1	1	0	0	0	0
6	1	1	0	0	0	1	1
5	1	0	1	0	1	0	2
4	1	0	0	0	1	1	3
3	0	1	1	1	0	0	4
2	0	1	0	1	0	1	5
1	0	0	1	1	1	0	6
0	0	0	0	1	1	1	7

Fig. 3.18 shows the waveforms associated with 3-bit down counter.

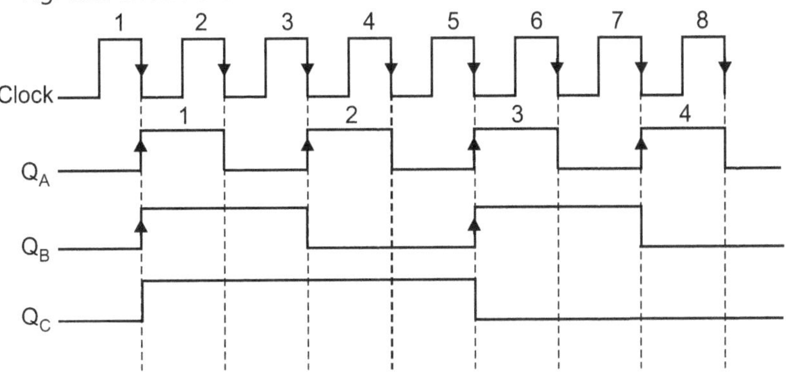

Fig. 3.18 : Waveforms of 3-bit down counter

- Thus, the count sequence - up or down depends upon how the flip-flops are connected. If we connect Q output to the clock input of next, it becomes up counter whereas if we connect $\overline{Q}$ to clock input of next, it becomes down counter.

3.12 Up/Down Counter

- In previous section, we have studied separate circuits for up counter and down counter.

- Using some gating circuitry, it is possible to connect either Q or $\overline{Q}$ at the clock input of next flip-flop and hence a single circuit can be used as either up counter or down counter. Such circuit which can be used as up counter or down counter is known as *up/down counter*.
  Fig. 3.19 shows the circuit for 3-bit asynchronous up/down counter.

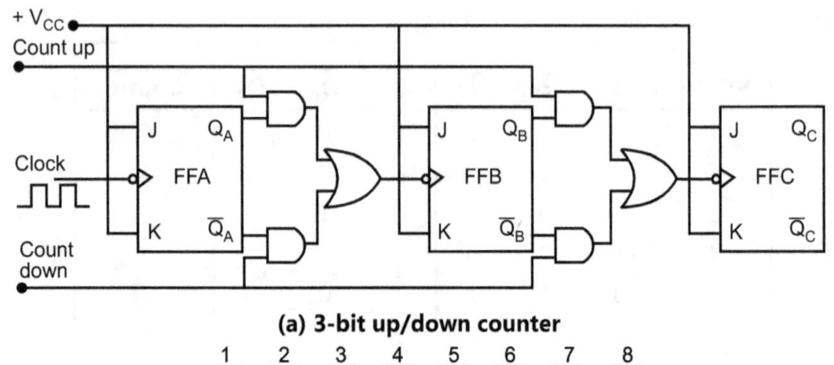

(a) 3-bit up/down counter

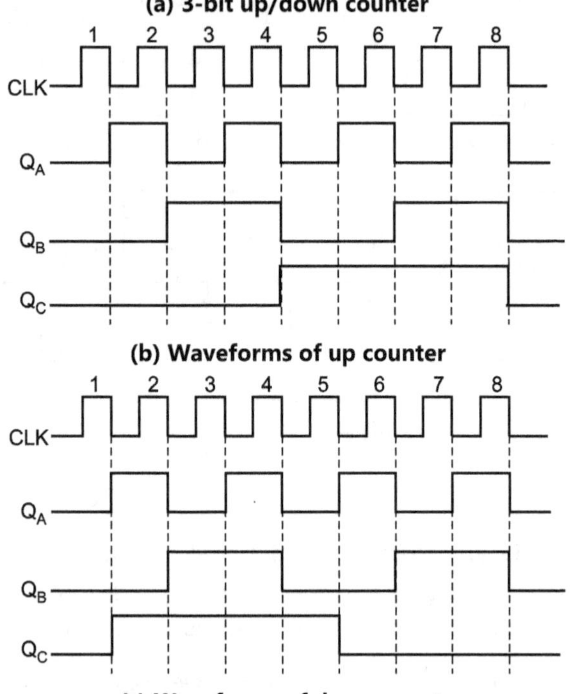

(b) Waveforms of up counter

(c) Waveforms of down counter

Fig. 3.19

- When the count up is held high and count down low, Q output of previous flip-flop gets connected to the clock input of next flip-flop and the circuit will behave as the up counter.

- Whereas when count up is held low and count down high, $\bar{Q}$ output of previous flip-flop gets connected to the clock input of next flip-flop and the circuit behaves as the down counter and causes the counter to progress through the count down sequence.

3.13 Modulus Counter

- The binary ripple counter can be constructed using clocked J-K flip-flop. The total number of counts or discrete states through which the counter can progress is given by 2^n, where n is the total number of flip-flops. Then it is said to have natural count of 2^n.

- A basic binary counter consisting of three flip-flop counts through eight discrete states is said to have natural count of 8.

- The 4-bit binary counter contains 4 flip-flop counts through 16 states. Thus, we can construct counters which count through 2, 4, 8, 16 ... etc. states by using proper number of flip-flops.

- A 3 flip-flop counter is often referred to as a modulus 8 counter since it has 8 states. The modulus of a counter is the total number of states through which the counter can progress.

- It is often desirable to construct counters which have moduli other than 2, 4, 8, ...etc. e.g. we may need to construct a counter having modulus 3 or 5 or 7.

- A smaller modulus counter can always be constructed from a larger modulus counter by skipping states. Such counters are said to have a modified count.

- The number of flip-flops required to construct desired counter is determined by choosing the lowest natural count which is greater than the desired modified count.

- Mod 7 counter requires 3 flip-flops since lowest natural count greater than 7 is 8 which requires 3 flip-flops. A mod 9 counter requires at least 4 flip-flops since 16 is the lowest natural count greater than 9 and requires 4 flip-flops.

- Generally, method used to skip count is to feedback a signal from some flip-flop to some previous flip-flop. e.g. consider a designing of mod 7 counter.

- It requires 3 flip-flops and since only one count is to be skipped, it would be convenient to let the counter advance through its natural sequence and then reset it one count early.

- Fig. 3.20 shows the logic diagram, truth table and waveforms of mod 7 counter.

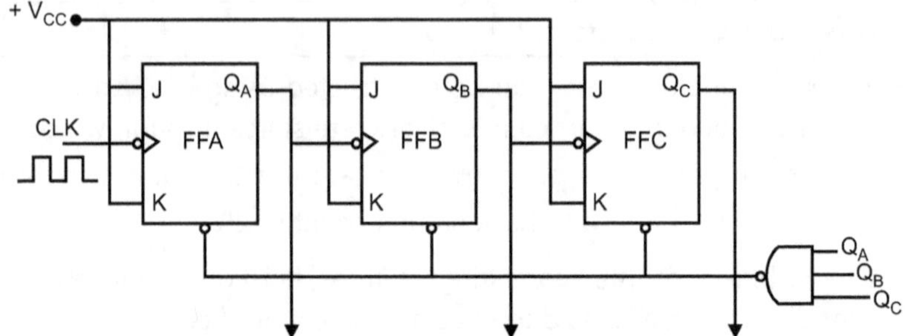

(a) Logic diagram

Q_C	Q_B	Q_A	Count
0	0	0	0
0	0	1	1
0	1	0	2
0	1	1	3
1	0	0	4
1	0	1	5
1	1	0	6
0	0	0	Reset

(b) Truth table

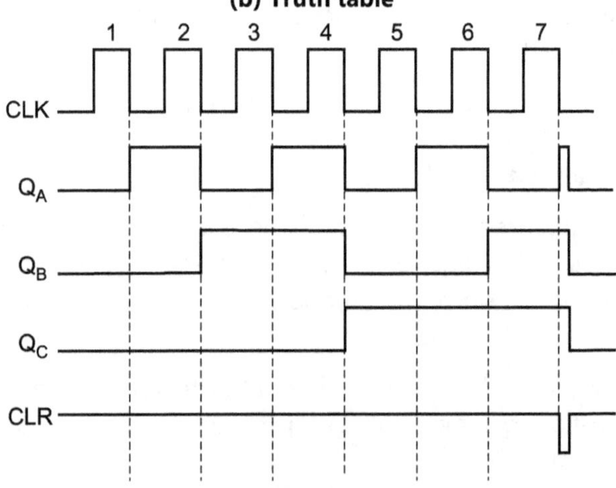

(c) Waveforms

Fig. 3.20

- From the truth table of natural 3-bit counter, we know that, count 7 (111) occurs only once. If Q_A, Q_B and Q_C outputs are connected to 3 inputs of 3 input NAND gate, the output of NAND gate goes low only when $Q_A = Q_B = Q_C = 1$.

- If the output of the NAND gate is connected to the CLR input of each flip-flop all the flip-flops will be cleared to zero's immediately as the counter advances to count 7. The counter remains in state 7 only momentarily, therefore, it is a mod 7 counter.

3.14 Decade Counter

- The counter which progresses through 10 different states is known as a *decade counter*. It is also known as *mod 10 counter*.
- Mod 10 counter can be constructed by skipping the states from natural 4-bit counter.
- The lowest natural count greater than 10 is 16 which requires 4 flip-flops and hence mod 10 counter also requires 4 flip-flops.
- Since counters with 4 flip-flops are capable of counting 16 discrete states, therefore, mod 10 counter can be obtained by skipping 6 states. This is done by using CLR pin of the flip-flop. Fig. 3.21 shows the mod 10 counter.

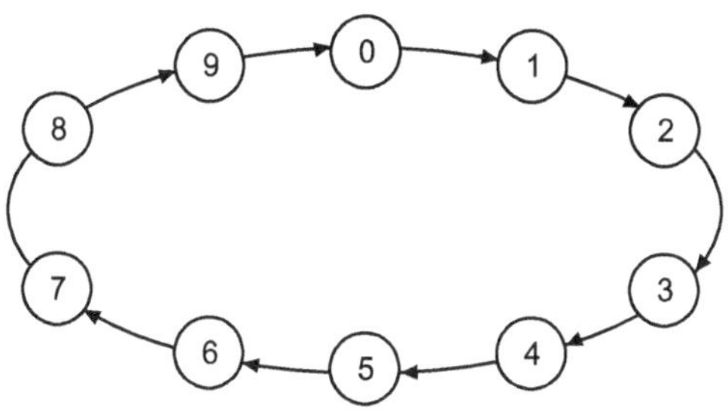

(a) State diagram

Count	Q_D	Q_C	Q_B	Q_A
0	0	0	0	0
1	0	0	0	1
2	0	0	1	0
3	0	0	1	1
4	0	1	0	0
5	0	1	0	1
6	0	1	1	0
7	0	1	1	1
8	1	0	0	0
9	1	0	0	1
Reset	0	0	0	0

(b) Truth table

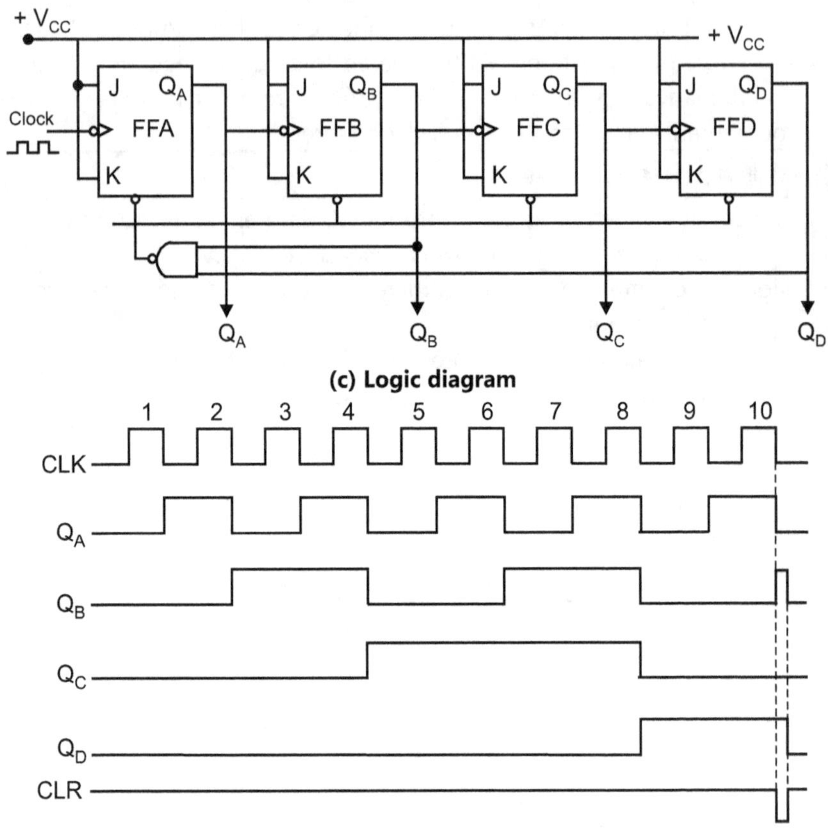

(c) Logic diagram

(d) Timing diagram

Fig. 3.21 : Mod 10 or decade counter

- The counter will start counting from 0000, when the counting reaches 1010 state clear pulse is generated by the NAND gate, then it will reset all flip-flops making content of counter 0000 and hence the circuit becomes mod 10 or decade counter.

3.15 IC 7490 as a Decade Counter

- Since in our day-to-day life, we are using decimal number system, so the decade counters which count from 0 to 9 have frequent applications.

- As seen, decade counters can be constructed using 4 flip-flops with feedback connections using discrete circuit. However, to reduce the burden on the designer, decade counters are available in IC form.

- IC 7490 is a decade counter MSI chip available from TTL family. IC is available in 14 pin DIP package. Fig. 3.22 shows the pin configuration, truth table and block diagram of IC 7490.

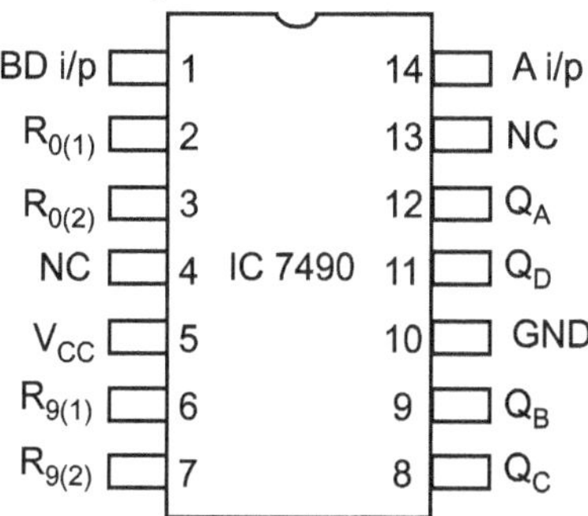

(a) Pin configuration

Count	Q_D	Q_C	Q_B	Q_A
0	0	0	0	0
1	0	0	0	1
2	0	0	1	0
3	0	0	1	1
4	0	1	0	0
5	0	1	0	1
6	0	1	1	0
7	0	1	1	1
8	1	0	0	0
9	1	0	0	1
Reset	0	0	0	0

(b) Truth table

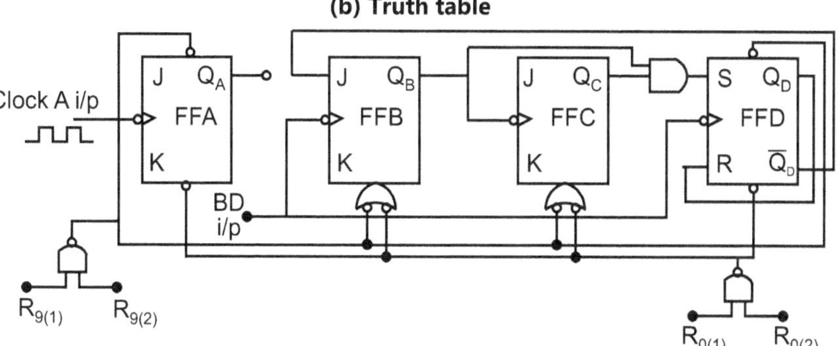

(c) Internal block diagram

Reset inputs				Outputs			
$R_{0(1)}$	$R_{0(2)}$	$R_{9(1)}$	$R_{9(2)}$	Q_D	Q_C	Q_B	Q_A
H	H	L	X	L	L	L	L
H	H	X	L	L	L	L	L
X	X	H	H	H	L	L	H
X	L	X	L	count			
L	X	L	X	count			
L	X	X	L	count			
X	L	L	X	count			

$$H \;=\; \text{High logic}$$
$$L \;=\; \text{Low logic}$$
$$X \;=\; \text{Irrelevant}$$

(d) Reset / Count function table

Fig. 3.22

- Internally IC 7490 is divided into two parts flip-flop A and flip-flops B, C and D. Flip-flop A functions as mod 2 and flip-flops B, C and D are connected in mod 5 configuration. IC 7490 is a versatile IC that can be used as a mod N counter with N ranging from 2 to 10.

- IC contains 3 J-K and one RS flip-flop, having 2 clock inputs A and BD, and four outputs Q_A, Q_B, Q_C and Q_D. It has 2 reset inputs $R_{0(1)}$ and $R_{0(2)}$ to reset or to clear the counter and two preset inputs $R_{9(1)}$ and $R_{9(2)}$ to set the counter to 1001. These inputs are also known as reset to 9 (R_9) inputs.

- Using single IC we can divide the frequency by either 2, 3, 4, 6, 7, 8, 9 or 10 and hence IC 7490 is also known as *scalar*. If we want modulus higher than 10 or frequency division greater than 10, we can cascade the IC 7490.

- The cascading is done in such a way that the terminal count of the previous counter stage is given as clock to the next stage. If we use 3 IC 7490, it can count from 000 to 999.

3.16 Synchronous Counters

- The ripple counters are simple to build, but they cannot be operated by high frequency signal.

- Since each flip-flop has a delay time, in a ripple counter these delay times are additive and the total settling time for the counter is approximately the delay time of the total number of flip-flops.

- If each flip-flop has a propagation delay of 10 ns, overall delay for 4-bit counter becomes 40 ns. This speed limitation can be overcome by the use of synchronous or parallel counter.

- Here every flip-flop is triggered by the clock and hence they all make their transitions simultaneously and hence known as *synchronous* or *parallel counter*. Fig. 3.23 shows the logic diagram, truth table and the timing diagram of mod 8 synchronous counter.

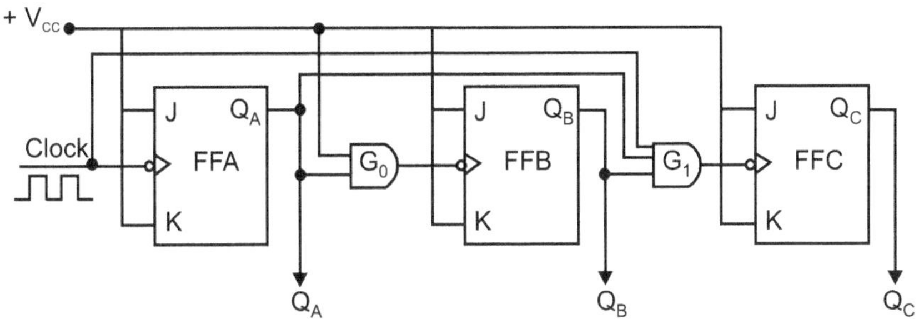

(a) Logic diagram

C	B	A	Count
0	0	0	0
0	0	1	1
0	1	0	2
0	1	1	3
1	0	0	4
1	0	1	5
1	1	0	6
1	1	1	7
0	0	0	0

(b) Truth table

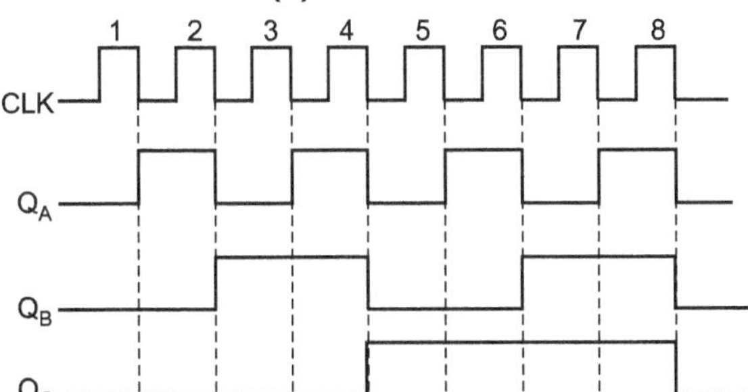

(c) Timing diagram

Fig. 3.23

- The parallel counter is constructed using J-K flip-flop with negative edge triggering. The flip-flop A changes the state with each negative transition at the clock input.
- The output of AND gate G_0 goes high whenever the clock is high and Q_A is high. Thus, flip-flop B changes the state whenever Q_A is high and clock goes negative.
- The output of AND gate G_1 goes high when the clock is high and both Q_A and Q_B is high. Thus, flip-flop C changes its state when both Q_A and Q_B is high and clock goes negative.
- From the Fig. 3.23 (c) we see that Q_A changes the state after every clock pulse, Q_B changes the state with every other clock pulse and flip-flop C changes the state with every fourth clock pulse and hence the circuit represents mod 8 counter.
- The parallel counter shown in Fig. 3.23 is the basic building block for building counters of other moduli. If we want to build mod 16 parallel counter, we require 4 flip-flops and one more 4 input AND gate.
- Whereas if we want to construct a mod 7 counter, it is necessary to eliminate the one state from the natural count sequence of mod 8 counter.
- One way to skip the last state is to use a feedback connection as we have seen previously and generate the clear pulse to clear the flip-flop earlier. We can also skip the first state.
- If by some means we prevent flip-flop A being reset during the transition from count 7 to count 0, without affecting the operations of B and C, the counter would progress from count 7 to count 1. Thus, count 0 would be skipped and a mod 7 counter would be formed.
- Fig. 3.24 shows the logic diagram, truth table and waveforms of the mod 7 synchronous counter.

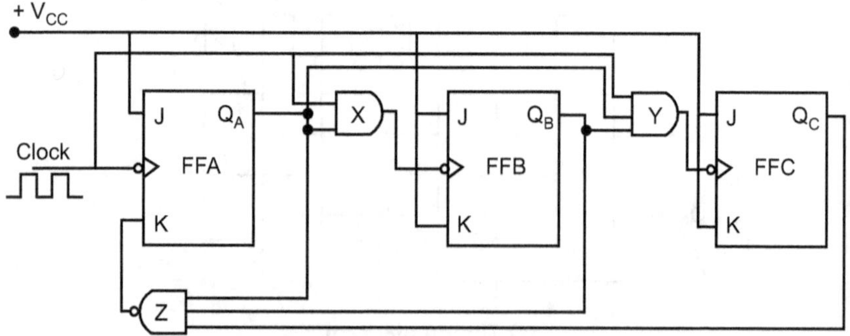

(a) Logic diagram of mod 7 synchronous counter

C	B	A	Count
0	0	1	1
0	1	0	2
0	1	1	3
1	0	0	4
1	0	1	5
1	1	0	6
1	1	1	7
0	0	1	1

(b) Truth table

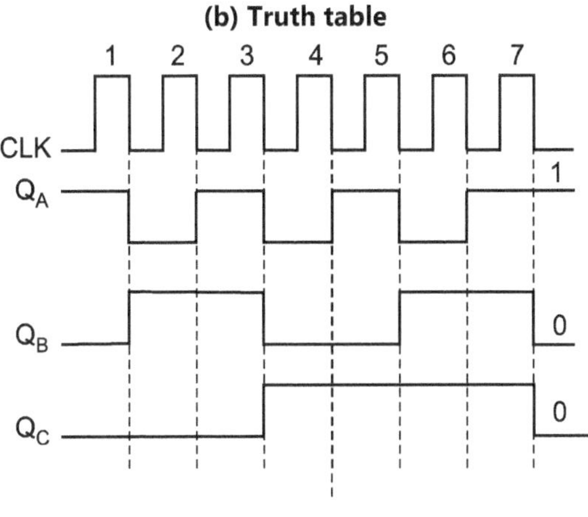

(c) Waveforms

Fig. 3.24

- For the logic diagram shown in above figure, counter advances by its natural sequence upto count 7. During the count, output of all flip-flops is high and since they are connected to input of NAND, it will generate low output which in turn holds K low.

- When the next clock pulse is applied, since J = 1, K = 0 which does not changes the stage of flip-flop A, and hence output of the counter goes from 111 to 001 state skipping 000 state. In turn, the circuit becomes the mod 7 counter.

- Thus, basically two types of counters can be designed and are also available in the market. Depending upon the requirement and the feasibility, user has to select the type of the counter.

- The points of difference between synchronous and asynchronous counters are as given below.

Difference between Synchronous and Asynchronous Counter :

Synchronous (parallel) counter	Asynchronous (ripple) counter
1. Every flip-flop is triggered by the clock in synchronism. Therefore, they make their transitions simultaneously.	1. The clock pulse is applied to the clock input of first flip-flop only. Each flip-flop is triggered by the output of previous flip-flop.
2. Setting time is less and is equal to the delay time of the single flip-flop.	2. Setting time is more and is equal to the sum of propagation delay times of total number of flip-flops.
3. Speed of operation is more and can be activated by high clock frequency.	3. Speed of operation is limited, the highest clock frequency depends on overall delay in the system.
4. Circuit is complex, more hardware is required and hence cost is more.	4. Simple circuit, easy to build, less hardware is required and hence cost is less.

Solved Examples

Example 3.1 : *In a 4 stage ripple counter, if the propagation delay of flip-flops is 20 ns and if the pulse width of the strobe is 20 ns, find the maximum frequency at which counter can operate reliably.*

Solution : The maximum frequency is,

$$f_{max} = \frac{1}{(4 \times 20 + 20)} \, ns = \frac{10^3}{4 \times 20 + 20} \, MHz$$

$$= \frac{10^3}{100} \, MHz = \frac{10^3}{10^2} \, MHz = 10 \, MHz$$

Example 3.2 : *Design a divide by 2 counter using IC 7490. Write its truth table and draw the waveforms.*

Solution :

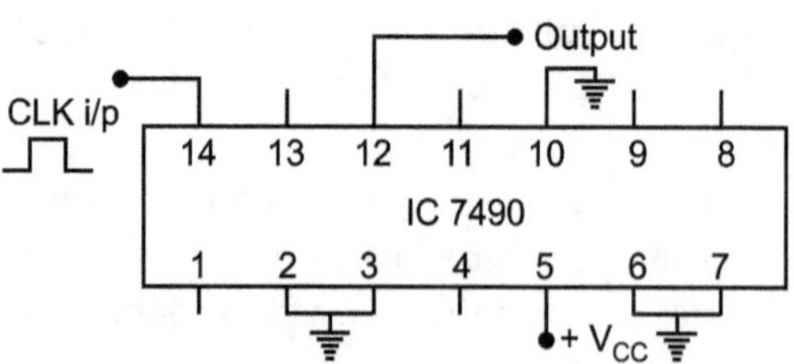

(a) Divide by 2 counter

CLK input	Output Q_A
0	0
1	1
2	0
3	1

(b) Truth table

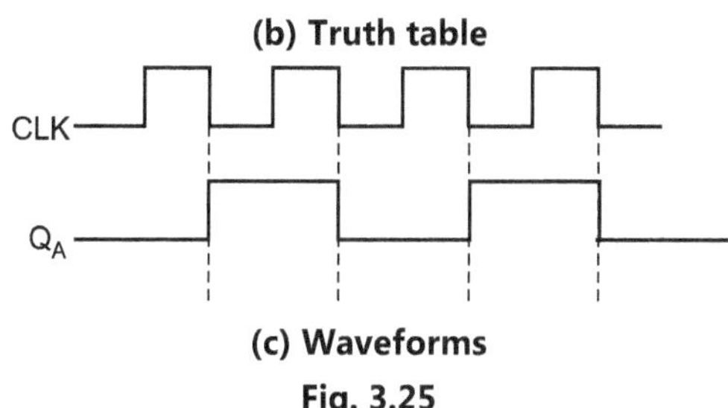

(c) Waveforms

Fig. 3.25

Example 3.3 : *Construct MOD7 counters using IC 7490.*

Solution : Modulus of a counter means number of stages through which counter progress. Mod-7 counter should have 7 different stages. IC 7490 has 10 states so to obtain mod 7 from IC 7490 we have to skip 3 states that can be done using the pins Rq(1) and Rq(2). When Rq(1) and Rq(2) becomes '1' counter reset to 1001, after that, it automatically goes to 0000 when next pulse is applied.

Table shows the count sequence and the circuit diagram for mod 7 using IC 7490.

Truth table

D	C	B	A	States	
0	0	0	0	1	
0	0	0	1	2	
0	0	1	0	3	
0	0	1	1	4	
0	1	0	0	5	
0	1	1	0	6	
0	1	1	1	7	Skip this state by connecting Q_B and Q_C to Rq(1) and Rq(2)
1	0	0	1	7	
0	0	0	0		← Reset

Circuit diagram :

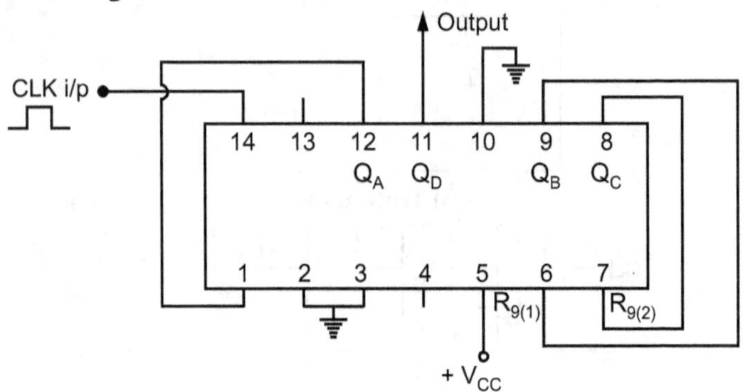

Fig. 3.26

Example 3.4 : *Construct the MOD3 counter using decade counter.*

Solution :

Truth table

Circuit	Output			
	D	**C**	**B**	**A**
0	0	0	0	0
1	0	0	0	1
2	0	0	1	0
3	0	0	1	1 ← Counter should reset

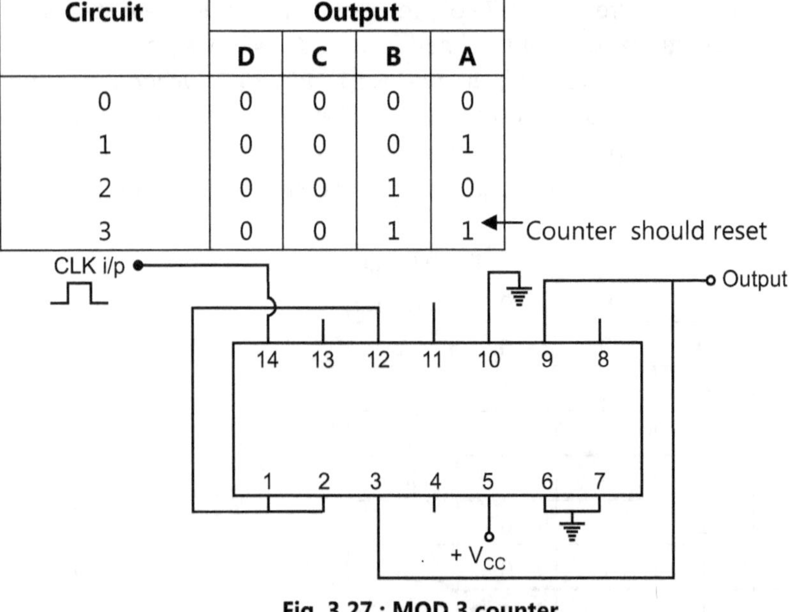

Fig. 3.27 : MOD 3 counter

Example 3.5 : *How many flip-flops are required to construct each of the following counters : mod 5, mod 10 ?*

Solution : The number of flip-flops required is determined by choosing the lowest natural count which is just greater than the desired modified count.

(i)   Mod 5 : The lowest natural count greater than 5 is 8. 3 flip-flops are required to construct counter 8. Therefore, mod 5 counter requires 4 flip-flops.

(ii) Mod 10 : The lowest natural count greater than 10 is 16. Therefore 4 flip-flops are required to construct counter mod 10.

Example 3.6 : *A 5-bit ripple counter begins with 00000 state. What will be the state of the counter after 100 input pulses ?*

Solution : For 5-bit counter, the maximum count is $= 2^n - 1 = 2^5 - 1 = 32 - 1 = 31$.

Therefore, counter resets on 32, 64, 96 and 128 input pulses. At 100th clock pulse, count will be equivalent to $100 - 96 = 4$ which is 00100. Therefore, state of the counter after 100 input clock pulses = 00100.

Example 3.7 : *Find the frequency of MSB output waveform in a 4-bit binary counter, if the square wave clock frequency is 12 MHz.*

Solution : In a 4-bit binary counter, the MSB output frequency can be obtained by dividing the clock frequency by 16. Therefore, MSB output waveform will have frequency $= \dfrac{12 \text{ MHz}}{16} = \dfrac{12 \times 10^3 \text{ kHz}}{16} = 750 \text{ kHz}$

Example 3.8 : *The frequency of a square wave is to be reduced from 100 kHz to 1 Hz. How can this be done using decade counters ? Explain using block diagram.*

Solution : We know decade counters divide the input clock frequency by 10, so we can have 5 stages of the decade counter as shown in Fig. 3.28.

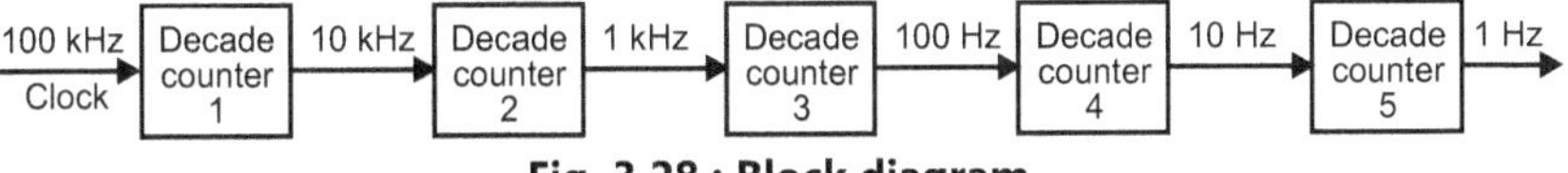

Fig. 3.28 : Block diagram

3.17 Introduction to Shift Registers

- A flip-flop is basically a one-bit storage cell. The group of flip-flop that is used to store binary number is called a *register*.
- One flip-flop is required to store one-bit and hence a register used to store 8-bit binary number must contain 8 flip-flops.
- The flip-flops must be connected such that the binary number can be entered (shifted) into the register and can be shifted out. A group of flip-flops connected to provide these functions is called a *shift register*.

- A register is normally used for storing and shifting data (1s and 0s) entered into it from an external source and possesses no characteristic internal sequence of states like that of a counter.
- Shift register is capable of shifting its binary information either to the right or to the left. The data in the register can be loaded serially or in parallel and can be shifted out either serially or in parallel, which leads to four basic types of registers.
- Fig. 3.29 shows symbolically the types of data movement in shift register operation. The block represents an arbitrary four-bit register and the arrow indicates the direction and type of data movement.

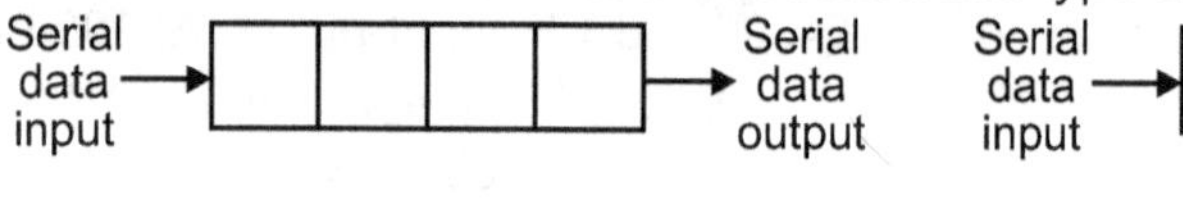

(a) Serial-in serial-out shift register

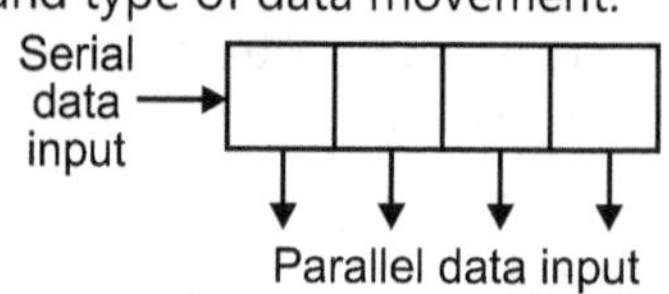

(b) Serial-in parallel-out shift register

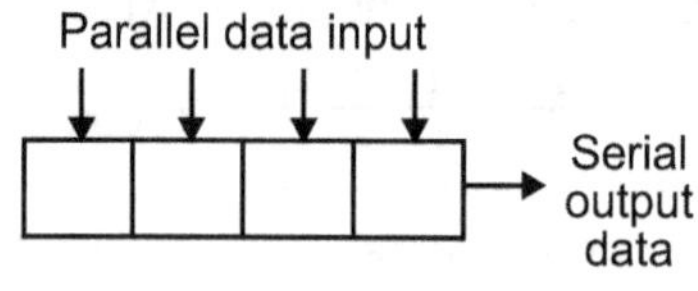

(c) Parallel-in serial-out shift register

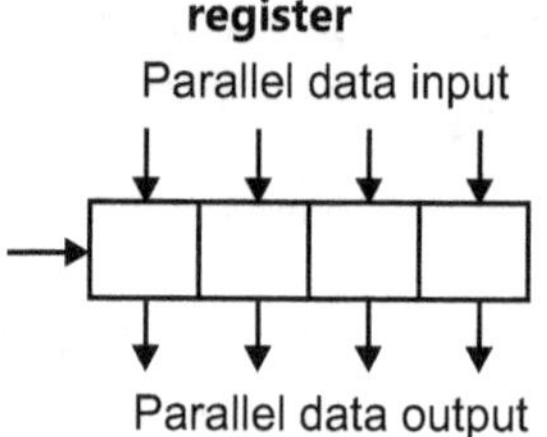

(d) Parallel-in parallel-out shift register

Fig. 3.29

3.18 Serial In Serial Out Shift Registers (SISO)

- This type of shift register accepts digital data serially, that is, one-bit at a time on a single line. It produces the stored information on its output also in serial form.
- Serial in serial out shift registers can be implemented either by using J-K or D flip-flop. Fig. 3.30 shows a 4-bit serial in serial out shift register using D flip-flop.
- Initially, all flip-flops are reset then the data to be entered can be applied to the D input starting with LSB or MSB.
- Let the data to be entered is 1001 beginning with LSB. The 1 is put on to the data input line, making D = 1 for FF_A. When the 1st clock pulse is applied, FF_A is set thus storing 1.

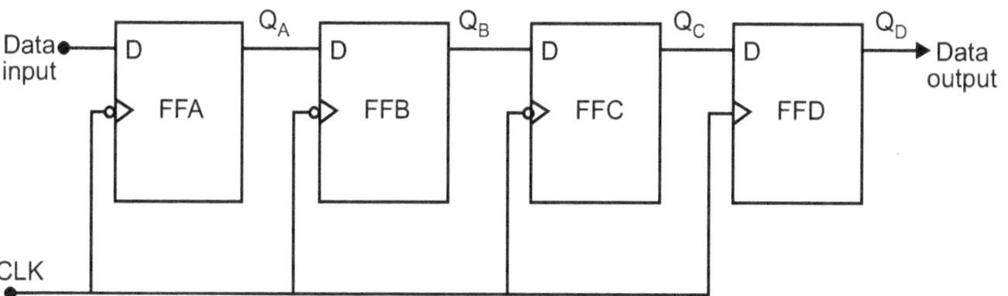

Fig. 3.30 : Serial in serial out shift register

- Next, 0 is applied to the data input, making D = 0 for FF_A and D = 1 for FF_B since Q_A is connected to D of FF_B. When the second clock pulse is applied, the 0 on the data input is shifted into FF_A and 1 that was in FF_A is shifted into FF_B.

- The next 0 in the binary number is now put on to the data line and a clock pulse is applied. The 0 is entered into FF_A. The 0 stored in FF_A is shifted into FF_B and 1 stored in FF_B is shifted into FF_C.

- The last bit in the binary number is 1, which is now applied to the data input and a clock pulse is applied. This time the 1 is entered into FF_A, the 0 stored in FF_A is shifted into FF_B, the 0 stored in FF_B is shifted into FF_C and the 1 stored in FF_C is shifted into FF_D. This completes the serial entry of the four-bit number into the shift register.

  Thus, before the application of the clock pulse,

  $$Q_A \, Q_B \, Q_C \, Q_D \quad = \quad 0000$$

  After one clock pulse, it becomes 1000

  $$Q_A \, Q_B \, Q_C \, Q_D \quad = \quad 0100 \qquad \text{after two clock pulses}$$
  $$= \quad 0010 \qquad \text{after three clock pulses}$$
  $$= \quad 1001 \qquad \text{after four clock pulses}$$

  Thus, to shift 4-bit data into 4 flip-flop requires four clock pulses.

- If we want to get the data out of the register, they can be shifted out serially and taken out at the Q_D output.

- After CLK_4 in the data entry operation described above the right most 1 in the number appears on the Q_D outputs.

- When CLK_5 is applied, the second bit appears on the Q_D output. CLK_6 shifts the third bit to the output and CLK_7 shifts the fourth bit to the output. When the stored bits are being shifted out, a new four-bit number can be shifted in.

- The shift registers are also available in IC forms. The 7491A is an example of an IC serial in serial out shift register.

3.18.1 The 7491A Eight Bit Shift Register

- IC 7491A is an example of an IC serial in serial out shift register. It contains eight S-R flip-flops. Fig. 3.31 shows the logic diagram of the same. As seen, there are two gated 31 input lines A and B, for serial data entry.

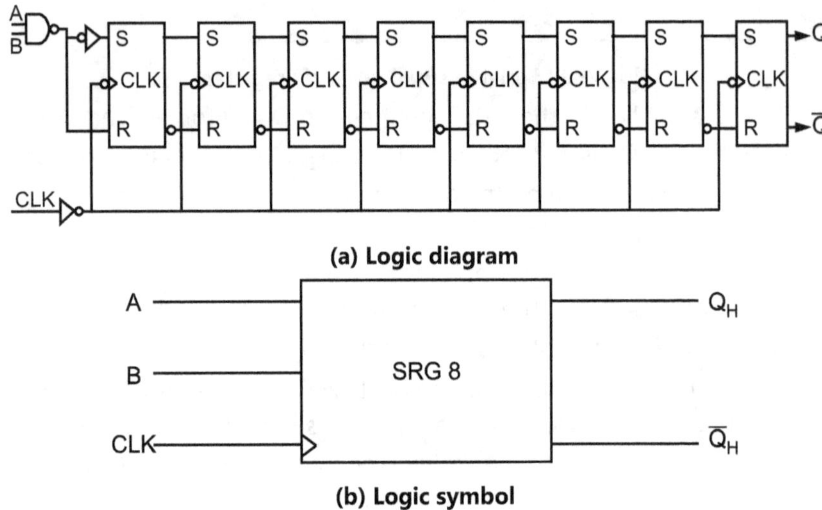

(a) Logic diagram

(b) Logic symbol

Fig. 3.31 : IC 7491A eight bit serial in serial out shift register

- The data can be entered either at A or B input. When data are entered at A input, B input must be high and vice versa. For user's convenience, manufacturer has provided two data inputs.

- Let data is applied at A input, during that time B is kept high. When applied data bit is 1, output of NAND gate is zero which makes $S = 1$ and $R = 0$.

- When the first clock pulse is applied, data is entered into the first shift register flip-flop, by the application of next successive pulses data propagates through shift register flip-flops after eighth pulse data is available at Q_H and its complement at $\overline{Q}_H$.

3.19 Serial In Parallel Out Shift Register (SIPO)

- In this type of register, data bits are shifted in serially but shifted out in parallel. In order to take parallel output, the output of each stage must be available. Once the data is stored, each bit appears on its respective output line and all bits are available simultaneously.

- Fig. 3.32 shows the 4-bit serial in parallel out shift register.

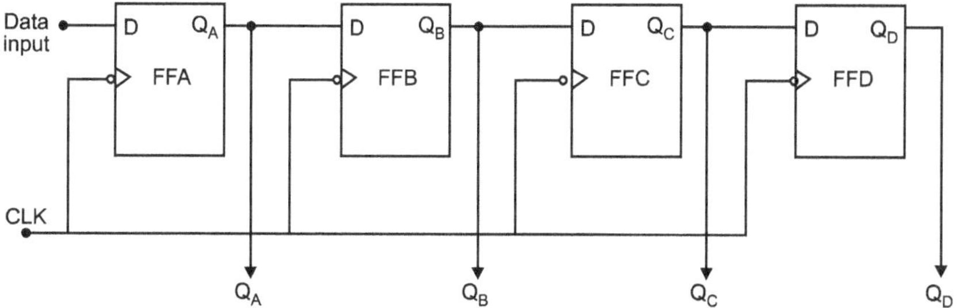

Fig. 3.32 : Serial in parallel out shift register

The IC 74164 is eight-bit serial in parallel out shift register.

3.19.1 IC 74164 Eight-Bit Serial in Parallel Out Shift Register

- The 74164 is an example of an IC shift register having serial in parallel out operation. The logic diagram is shown in the Fig. 3.33.

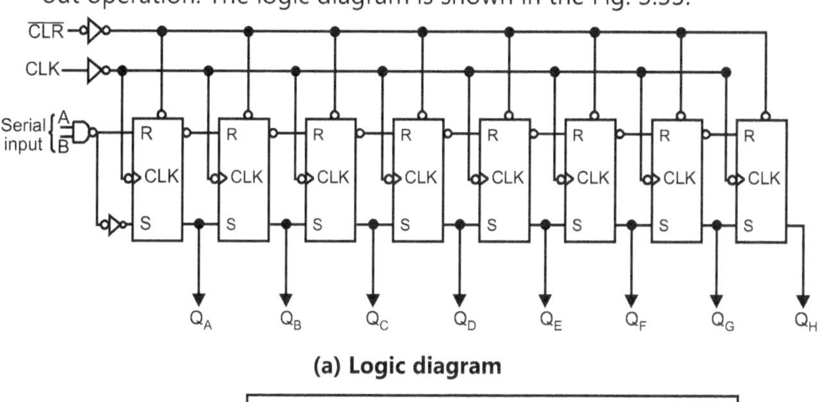

(a) Logic diagram

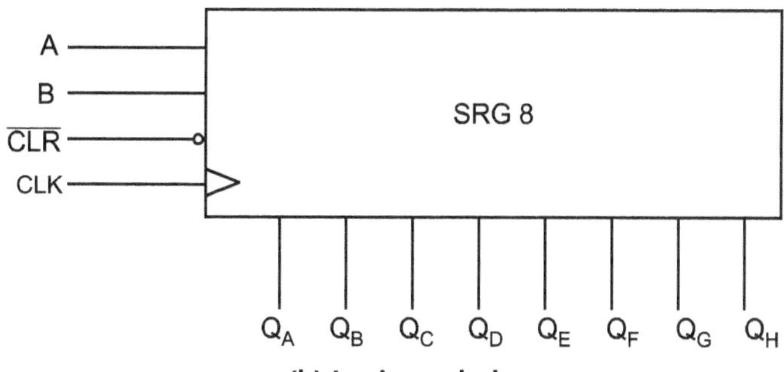

(b) Logic symbol

Fig. 3.33 : IC 74164 eight-bit serial in parallel out shift register

- From Fig. 3.33 we see that the device has two serial gated inputs A and B. It has active low clear input and eight outputs Q_A through Q_H.

- Data to be shifted in can be applied to either data input, provided the other data input is kept high. Initially, all flip-flops are cleared by applying active low clear pulse.
- Then suppose data is applied at serial input A and serial input B is kept high, at the end of first clock pulse first data bit gets entered into the first flip-flop and appears at Q_A. At the end of second clock pulse same bit gets shifted into the second flip-flop and also appeared at Q_B and at the same time new data bit gets entered into flip-flop A.
- In this way, 8-bit data gets entered into eight respective flip-flops and appears at the output at the end of eighth clock pulse.

3.20 Parallel In Serial Out Shift Register (PISO)

- In this case, data bits are entered simultaneously into their respective stages. Once the data is completely stored in the register, it can be shifted serially out, one-bit at a time using clock pulses.

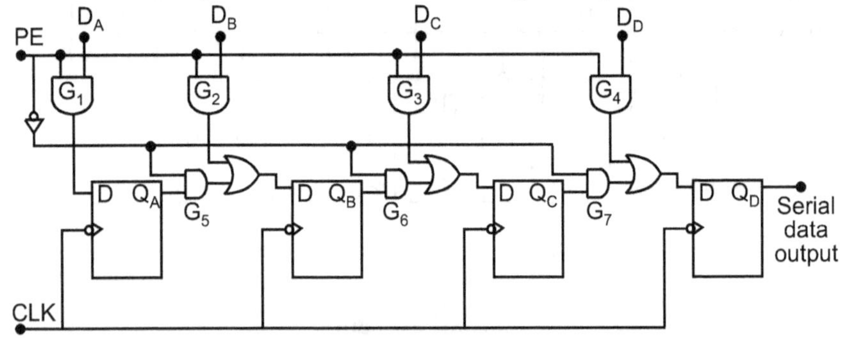

Fig. 3.34 : A four-bit parallel in serial out shift register

- Fig. 3.34 shows the four-bit parallel in serial out shift register. There are four data inputs D_A, D_B, D_C, D_D and a parallel enable line which allows four bits of data to be entered in parallel into the register.
- When PE line is high, gates G_1 through G_4 are enabled which allow each data bits to be applied, to the D inputs of respective flip-flops. When a clock pulse is applied, the flip-flops with a 1 data bit will set and those with 0 data bit will reset. Thus, it stores the data word in one clock cycle i.e. storing all four bits simultaneously.
- When the PE line is low, parallel data input gates G_1 through G_4 are disabled and shift enable gates G_5 through G_7 are enabled which allow data bits to shift from one stage to the next after each clock pulse.

- The IC 74165 is an example of parallel in serial out on a IC shift register.

3.20.1 IC 74165 Parallel Load Shift Register

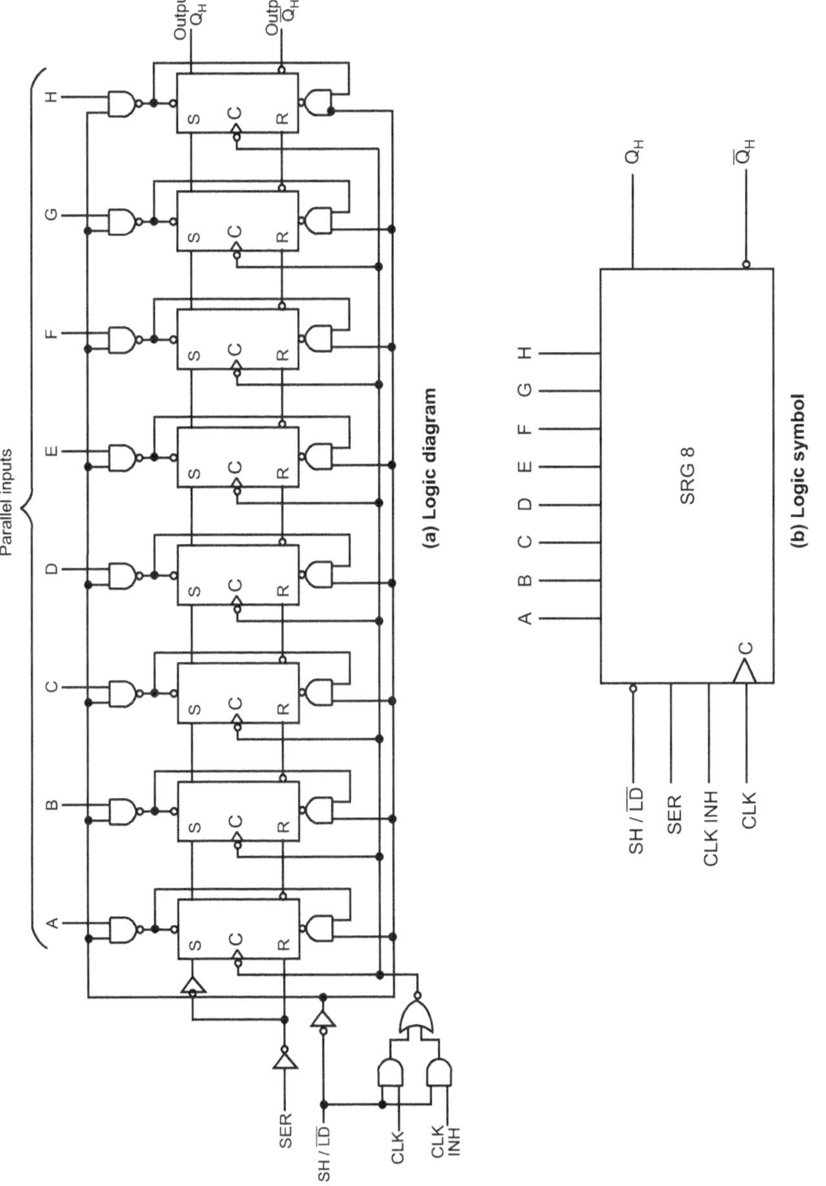

Fig. 3.35 : The 74165 eight-bit parallel load shift register

- IC 74165 is an example of IC shift register. It can be operated in parallel in serial out as well as serial in serial out operation.

- Fig. 3.35 shows the logic diagram and logic symbol of the IC 74165.

- As seen from the logic diagram, a low of the $\overline{\text{Shift/Load}}$ input enables all the NAND gates for parallel loading.

- When an input data bit is 1, the flip-flop is asynchronously SET by a low due to upper gate.

- When the input data bit is 0, the flip-flop is asynchronously RESET by a low out of the lower gate.

- During parallel loading operation, the clock is inhibited. A high on the $\overline{\text{SH/LD}}$ input enables the clock, causing the data in the register to shift right.

- The same IC can be used as serial in serial out shift register. For that purpose, data is entered serially in the SER input.

3.21 Parallel in Parallel Out Shift Register (PIPO)

- In parallel in parallel out shift register, data bits are entered simultaneously as well as taken out simultaneously. Fig. 3.36 shows such shift register.

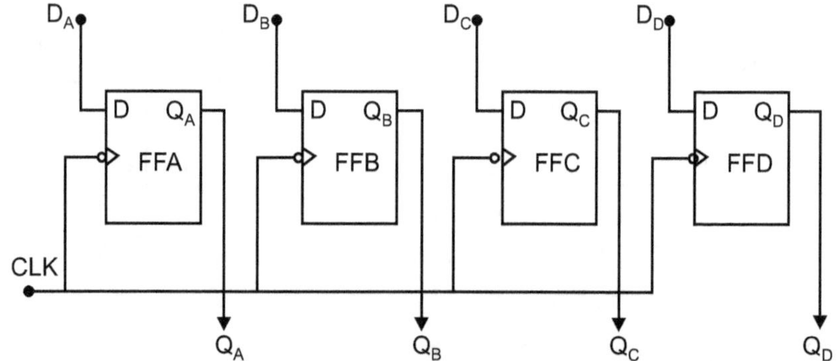

Fig. 3.36 : Parallel in parallel out shift register

- The data to be entered is applied to the data inputs of the respective flip-flop. After applying the clock pulse data gets stored in the respective flip-flop and also appears at the output from where it can be read simultaneously.

- The IC 74195/198 is an example of parallel in parallel out shift register.

3.21.1 IC 74195 Four-Bit Parallel In Parallel Out

(Parallel Access) Shift Register

- IC 74195 can be used for parallel in parallel out, serial in serial out and serial in parallel out operation. The logic diagram of the IC 74195 is shown in the Fig. 3.37.

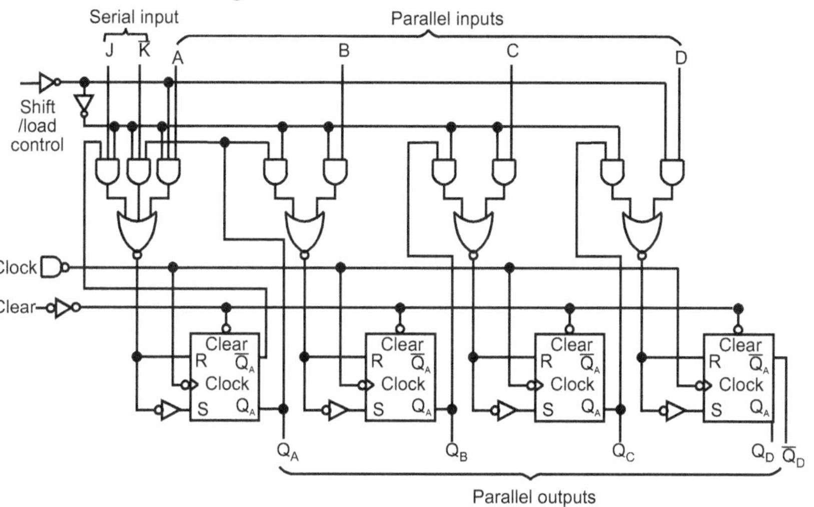

(a) Logic diagram

(b) Logic symbol

Fig. 3.37 : IC 74195 as shift register

- When shift/load input is low, the data on the parallel inputs are entered synchronously on the positive transition of the clock.

- When $SH/\overline{LH}$ input is high, stored data will be shifted right (Q_A to Q_D) synchronously with the clock. J and $\overline{K}$ are the serial data inputs to the first stage of the register Q_A. Q_D can be used for serial output data. The active low clear is asynchronous.

3.22 Universal Shift Register

- In the case of shift register, data bits are shifted in the flip-flop with the occurrence of clock pulses. A bidirectional shift register is one in which the data can be shifted to either left or right.

- The shift left or right is achieved by using gating logic that enables the transfer of data bit from one stage to the next stage to the right or to the left depending upon the level of the control line.

- A register is referred to as a universal shift register if it can be operated in all four possible modes and also as a bidirectional register. e.g. IC 74194.

- A 4-bit shift register using 4 flip-flops is shown in the Fig. 3.38. This circuit can be used in any of the four modes. The operation of this circuit is explained by assuming the 4-bit data 1001.

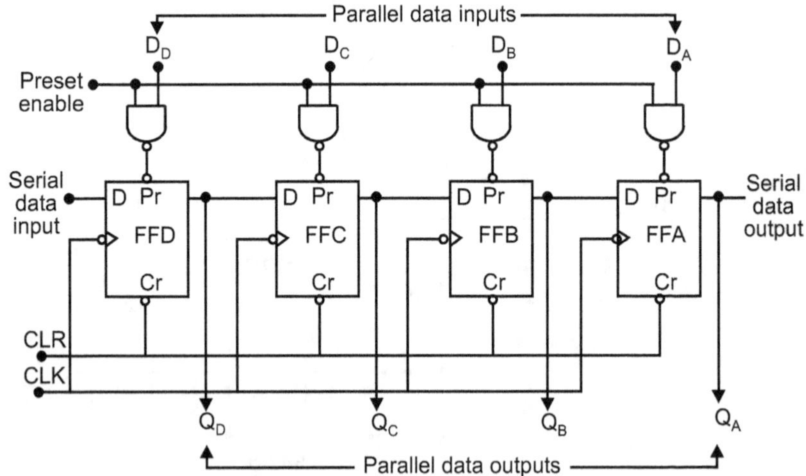

Fig. 3.38 : 4-bit shift register

1.   Serial input :

- Initially, all flip-flops are cleared by applying clear pulse i.e. making Cr = 0. Then for the normal operation, preset enable is made low which makes Pr = 1 and clear is also kept 1.

- The data word in the serial form is applied at the serial input. As described previously, data bits get entered one by one upon receiving next clock pulse.

- Four clock pulses are required to store 4-bit data and then serial data can be taken at serial output terminal after applying further clock pulses. The number of clock pulses required will be same as the number of bits.

- In the parallel form, the data is available at Q_D, Q_C, Q_B, Q_A and clock is not required for reading.

2. **Parallel input :**

- In this case, all data bits are entered simultaneously. For the logic circuit shown above, data can be entered in the parallel form making use of the preset inputs.

- First by applying the clear pulse all flip-flops are cleared. 1 is applied to the preset input and then data bits are applied.

- If we want to load 1001, respective flip-flop gets set, so data gets written into the register. The loaded data can be read simultaneously across Q_D, Q_C, Q_B, Q_A or can be read serially by applying clock pulses across serial output.

3.23 Bidirectional Register

- There are certain applications in which it is required to shift data either to the right or to the left.

- As we know, shifting binary number to the right by one-bit having LSB zero is equivalent to divide the number by two and with MSB zero shifting binary number to the left by one-bit is equivalent to multiply by two.

- Bidirectional shift registers can be used for this purpose. A 4-bit bidirectional shift register is shown in Fig. 3.39.

- When the mode control is held high M = 1, all G gates (AND Gates) are enabled and the data at D_R is shifted to the right when clock pulses are applied.

- On the other hand, when M = 0, G gates are disabled and G' gates are enabled which allow the data at D_L to be shifted to the left. Thus, the mode control decides the shift left or right operation.

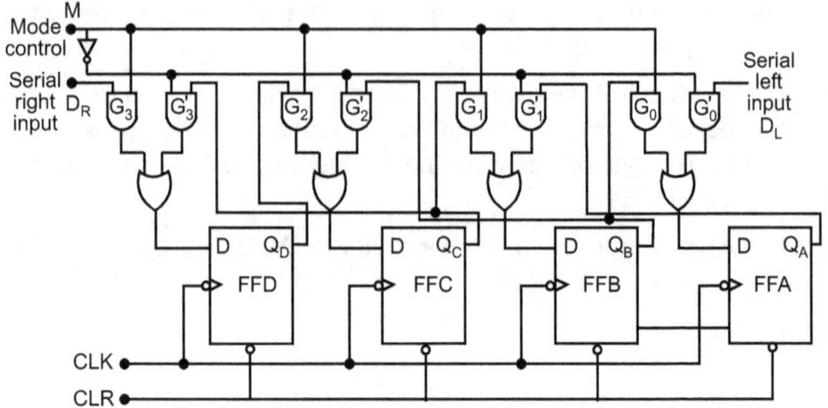

Fig. 3.39 : Bidirectional shift register

3.24 Universal Shift Register - IC 7495

- The shift register which can be connected in serial in serial out, serial in parallel out, parallel in serial out and parallel in parallel out configuration, as well as in which we can perform shift right and shift left operations is known as *universal shift register*.

- IC 7495 available in 14 pin DIP ceramic pack is an example of universal shift register. Fig. 3.40 (a) shows the pin configuration of IC 7495.

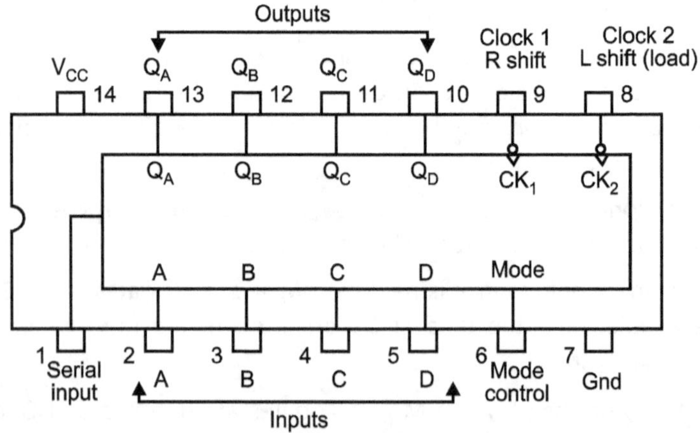

(a) Pin configuration of IC 7495

Inputs							Outputs			
Mode control	Clocks 2(L) 1(R)	Serial	Parallel				Q_A	Q_B	Q_C	Q_D
			A	B	C	D				
H	H	X	X	X	X	X	X	Q_{A0}	Q_{B0}	Q_{C0}	Q_{D0}
H	$\downarrow$	X	X	a	b	c	d	a	b	c	d
H	$\downarrow$	X	X	$Q_B\uparrow$	$Q_C\uparrow$	$Q_D\uparrow$	d	Q_{Bn}	Q_{Cn}	Q_{Dn}	d
L	L	H	X	X	X	X	X	Q_{A0}	Q_{B0}	Q_{C0}	Q_{D0}
L	X	$\downarrow$	H	X	X	X	X	H	Q_{An}	Q_{Bn}	Q_{Cn}
L	X	$\downarrow$	L	X	X	X	X	L	Q_{An}	Q_{Bn}	Q_{Cn}
$\uparrow$	L	L	X	X	X	X	X	Q_{A0}	Q_{B0}	Q_{C0}	Q_{D0}
$\downarrow$	L	L	X	X	X	X	X	Q_{A0}	Q_{B0}	Q_{C0}	Q_{D0}
$\downarrow$	L	H	X	X	X	X	X	Q_{A0}	Q_{B0}	Q_{C0}	Q_{D0}
$\uparrow$	H	L	X	X	X	X	X	Q_{A0}	Q_{B0}	Q_{C0}	Q_{D0}
$\uparrow$	H	H	X	X	X	X	X	Q_{A0}	Q_{B0}	Q_{C0}	Q_{D0}

H = high level, L = low level, X = irrelevant

$\downarrow$ = transition from high to low, $\uparrow$ = transition from low to high

(b) Function table

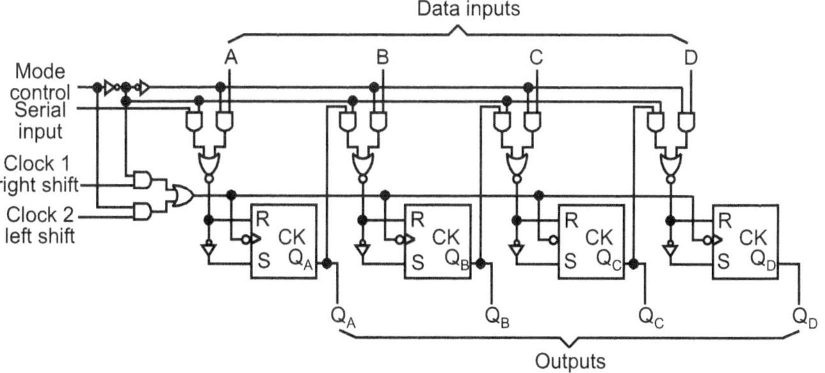

(c) Functional block diagram of IC 7495

Fig. 3.40

- Fig. 3.40 (c) shows the functional block diagram of IC 7495. Being a 4-bit shift register, it has 4-shift register flip-flop.

- It has mode control input, serial data input, clock 1 for right shift operation, clock 2 for left shift operation along with 4 data input and 4 data outputs.

Parallel in parallel out :

- Apply 4-bits of data and take the mode control input high. The data is loaded into the associated flip-flop. The loaded data appears at the output when clock 2 input is made high to low. During loading, the entry of serial data is inhibited.

Parallel in serial out :

- The loaded data can be shifted right (Q_A towards Q_D) and can be taken out at Q_D on high to low transition of clock 1 when the mode control is high. Next 3 clock pulses will be required to shift data out.

Serial in parallel out :

- When mode control is low, data applied at serial data input will be entered into flip-flop A after high to low transition of clock 1, at the end of 4 clock pulses data gets stored at 4 flip-flop and can be taken out parallels.

Serial in serial out :

- Keeping mode control low, data at serial input can be entered serially into the register, apply the data input mode clock 1 high to low, one bit gets entered, 4 clock pulses to enter 4-bit data and next 3 clock pulses to shift data out serially.

Shift left :

- In this mode, data moves from Q_D to Q_C, Q_C to Q_B, Q_B to Q_A. For this purpose, output of each flip-flop needs to be connected to the parallel input of previous flip-flop i.e. Q_D to input C, Q_C to input B, Q_B to input A and serial data is entered at input D. During this mode, control is held high. High to low transition of clock 2 shifts the data left.

- Thus, IC 7495 can be operated in all possible modes, being a versatile IC used in many applications.

3.25 Applications of Shift Register

- Shift registers are basically used to store data temporarily, besides this shift registers can be used in numerous applications.

1. **Time delay :**

- The serial in serial out shift registers can be used to provide a time delay. The time delay provided from input to the output is a function of both the number of stages in the register and the clock frequency.

- If t_d is the delay provided by the circuit then

$$t_d = N \times \frac{1}{f_c}$$

where N is the number of stages and f_c is the clock frequency.

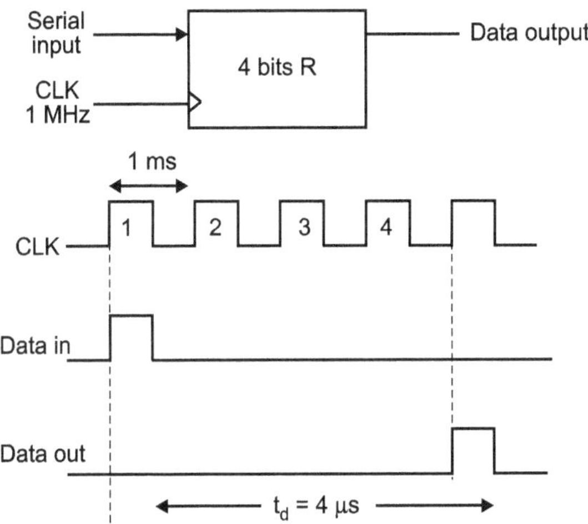

Fig. 3.41 : Shift register as a time delay device

- When a data is applied to the serial input on triggering edge of the clock pulse, it enters into the 1st stage. It is then shifted from stage to stage on each successive clock pulse after 4 clock pulses it appears at the output.

- If the clock frequency is 1 MHz, time delay provided is $4 \times 1\ \mu s = 4\ \mu s$. Time delay can be increased by cascading the shift register and decreasing the clock frequency.

2. Ring counter :

- If the final output of shift register is connected back to the serial input, a shift register can be used as a ring counter. In this, 1 continues to circulate through the counter and can be used to trigger certain devices periodically.

3. Parallel to serial converter :

- Parallel data transmission over a longer distance is quite impractical and hence for long distance data transmission data is converted from parallel to serial form and then transmitted over a pair or single line. Thus, at transmitting end, parallel in serial out shift registers are used.

4. Serial to parallel converter :

- Serially received data from a distant source cannot be fed directly to the computer. Since computers being a parallel device and hence at the receiving end serial in parallel out shift register are used to convert serially received data into parallel form and then fed to the computer for further processing.

- Now-a-days computers and microprocessor based systems are used to send and receive data. A dedicated hardware device known as UART which contains serial in parallel out shift registers and parallel in serial out shift registers are also used along with computers for long distance data communication.

Solved Examples

Example 3.9 : *If the clock frequency is 8 MHz, how long it will take to shift 8-bit number into a serial in serial out shift register ?*

Solution : Since the clock frequency is 8 MHz, hence clock period,

$$T = \frac{1}{f} = \frac{1}{8 \times 10^6} = 0.125 \ \mu s$$

To shift 8-bit data serially 8 clock pulses will be required and hence time for serial input will be $8 \times 0.125 \ \mu s = 1 \ \mu s$. To shift this data out of the circuit, 7 more clock pulses will be required, since after storing data 1-bit is available at the output of last flip-flop at the end of 8th clock pulse. So total pulses required will be $8 + 7 = 15$. So the total time will be $15 \times 0.125 \ \mu s = 1.875 \ \mu s$.

Example 3.10 : *Draw the waveforms to show the status of the 4-bit register in Fig. 3.42 for the specified data input. Assume that all flip-flops are cleared initially.*

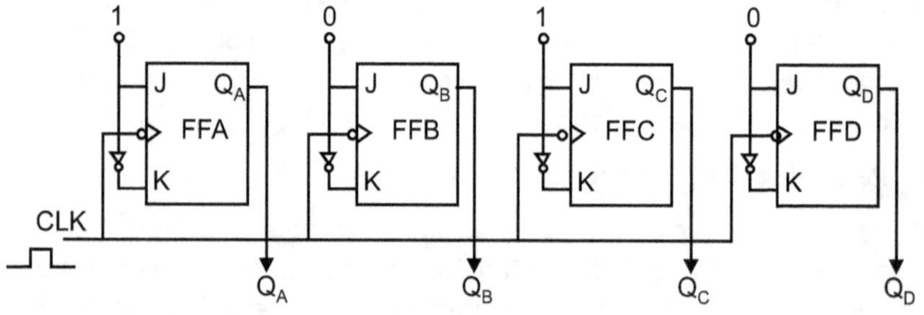

Fig. 3.42

Solution :

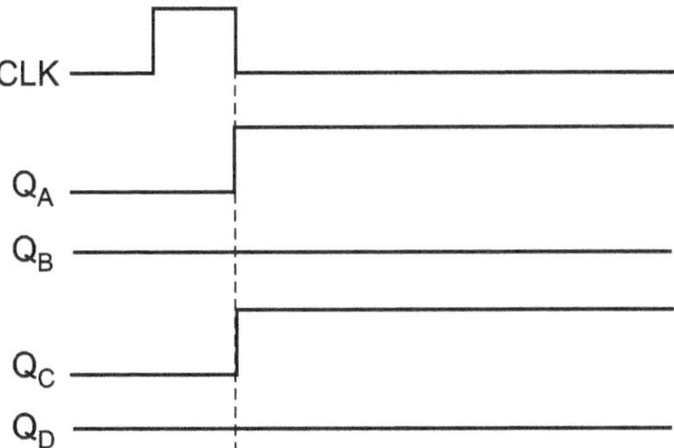

Fig. 3.43

3.26 Ring Counter

- A shift register is modified into a ring counter by connecting the serial output back to the serial input. It is known as the counter because it has specified sequence of states.

- A logic diagram for a four-bit ring counter is shown in Fig. 3.44.

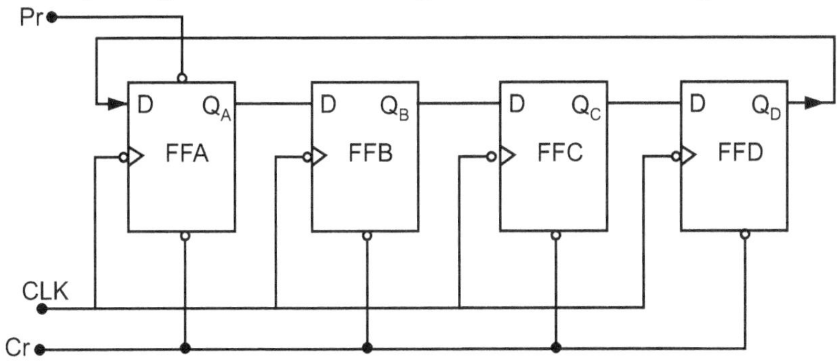

Fig. 3.44 : 4-bit ring counter

- Instead of counting in binary mode, the ring counter rotates the bit among the flip-flop. A logic diagram for 4-bit ring counter is shown in Fig. 3.44.

- To begin with, a 1 is preset into the 1st flip-flop and the remaining flip-flops are cleared. When 1st clock pulse is applied, 1 is shifted from Q_A to Q_B.

- When 2nd clock pulse is applied, 1 is further transferred from Q_B to Q_C and so on. 1 is always retained in the counter and simply shifted around the ring advancing one stage for each clock pulse. Table 3.5 shows the ring counter sequence.

Table 3.5 : Ring counter sequence

Clock pulse	Q_A	Q_B	Q_C	Q_D
0	1	0	0	0
1	0	1	0	0
2	0	0	1	0
3	0	0	0	1
4	1	0	0	0

Fig. 3.45 shows the waveforms for 4-bit ring counter.

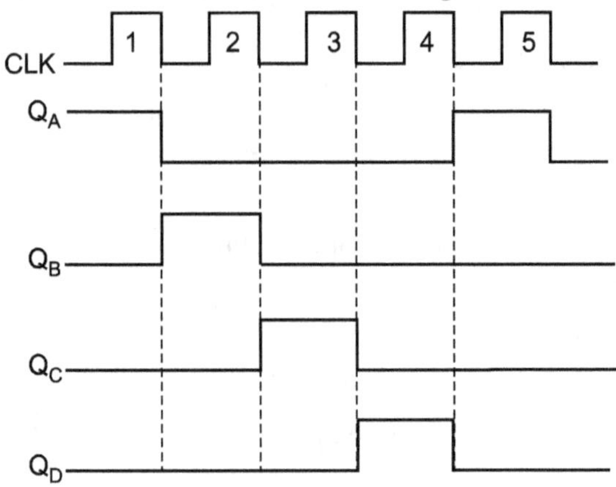

Fig. 3.45 : 4-bit ring counter waveforms

3.26.1 Uses of Ring Counters

- In case of a ring counter, the stored bit follows a circular path and hence it is very useful in timing sequence of digital operation.
  (i)   To control a sequence of operation.
  (ii) In stepper motor.
  (iii) In control state counters.
  (iv) As divide by N counter, where N = number of clock pulses.
  (v) To activate digital circuits in a computer at a right time and in the right sequence.

3.27 Event Counter

- Event counters are used to count and display the number of times an event occurs. For example, counting the number of people passing through a queue, or the number of boxes moving along a conveyor belt.

- Our event counter boards accept pulses from switches, relays, open or push button.

- It is an application of an counter. It basically has three parts, for converting events in to pulses, a counter and a circuits to display the count.

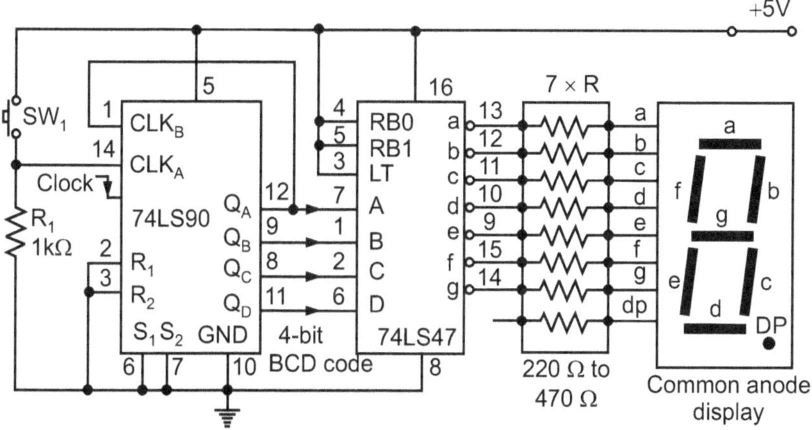

Fig. 3.46 : 4-bit BCD counter circuit

- Note that a 7-segment display is made of seven individual light emitting diodes to form the display. The best method of limiting the current through a seven segment display is to use a current limiting resistor in series with each of the seven LED's as shown in Fig. 3.46.

- The clock pulses are counted by the decade counter which are fed to display through decoder driver. In case of event counter, the events are converted into pulses. For such conversion, usually optical devices such as LDR, photodiodes are used.

Think Over It

1. Sequential circuit requires clock for their operation.

2. Presets and clears are normally active low.

3. Decade counter reset after counting nine i.e. 1001.

4. Synchronous counters ar faster than asynchronous counters.

Points to Remember

1. A flip-flop is a basic memory element used to store one bit of information.
2. A clocked flip-flop is a flip-flop whose state changes to the data inputs only when a clock pulse is present.
3. In a sequential system, it is required to set or reset the flip-flop in synchronism with clock pulses.
4. When $J = 1$, $K = 0$ sets the flip-flop when clock is applied.
5. When $J = 0$, $K = 1$ clears the flip-flop when clock is applied.
6. We can clear the flip-flop or set the flip-flop by addition of two more inputs to the flip-flop known as *preset* and *clear*.
7. A latch stores one bit of data. It is a D type of flip-flop.
8. 'Preset' input is used to set (make $Q = 1$) the flip-flop where as 'Clear' input is used to clear or reset (make $Q = 0$) the flip-flop.
9. In J-K flip-flop when $J = K = 1$ and clock is applied, the outputs go on complementing after every delay time of flip-flop as long as clock is present. This condition is known as *'Race around condition'*.
10. Master Slave flip-flop avoids or eliminates race around condition.
11. A register in which data gets shifted towards left or right when clock pulses are applied is known as a *shift register*.
12. The applications of shift registers are serial to parallel converter, ring counter, sequence generator, twisted-ring counter etc.
13. In synchronous (parallel) counter every flip-flop is triggered by the clock in synchronism. Therefore, they make their transitions simultaneously, whereas in asynchronous (ripple) counter the clock pulse is applied to the clock input of first flip-flop only, each flip-flop is triggered by the output of previous flip-flop.
14. In synchronous (parallel) counter, setting time is less and is equal to the delay time of the single flip-flop, whereas in asynchronous (ripple) counter, setting time is more and is equal to the sum of propagation delay times of total number of flip-flops.
15. Counter is basically constructed using flip-flops and is an example of sequential circuits.
16. A counter is used to count pulses and give the output in the binary form.
17. Many physical parameters or events can be converted into pulses and hence counter can be used to indicate temperature, vibrations etc. or to count events.

18. If the output of a counter is proportional to the binary equivalent of number of clock pulses applied at its input, it is known as *binary counter.*

19. The counter which progresses through 10 different states is known as a *decade counter.*

20. A synchronous counter is faster than an asynchronous counter.

21. The flip-flops connected such that the binary number can be entered (shifted) into the register and can be shifted out. A group of flip-flops connected to provide these functions is called a *shift register.*

22. Serial in serial out shift register accepts digital data serially that is one-bit at a time on a single line and output serially.

23. For serial in parallel out shift register data bits are shifted in serially but shifted out in parallel.

24. For parallel in serial out shift register data bits are entered simultaneously into their respective stages. Once the data is completely stored in the register, it can be shifted serially out, one-bit at a time using clock pulses.

25. In parallel in parallel out shift register, data bits are entered simultaneously as well as taken out simultaneously

26. Shift register can be modified into a counter by connecting the serial output back to the serial input. It is known as the counter because it exhibits a specified sequence of states.

27. Ring counters are used to control a sequence of operation. In stepper motor control, in control state counters, ring counters are used to activate digital circuits in a computer at a right time and in the right sequence.

28. A register is referred to as a universal shift register if it can be operated in all four possible modes.

Exercise

[A] Multiple Choice Questions:

1. Asynchronous counters are known as ....
   - (a) ripple counters
   - (b) multiple clock counters
   - (c) decade counters
   - (d) modulus counters

2. Asynchronous counter differs from an synchronous counter in ....
   - (a) the number of states in its sequence
   - (b) the method of clocking
   - (c) the type of flip-flops used
   - (d) the value of the modulus

3. A 4-bit binary counter has a maximum modulus of ....
    (a) 16   (b) 32   (c) 8   (d) 4
4. A BCD counter is an example of ....
    (a) a full-modulus counter   (b) a decade counter
    (c) a truncated-modulus counter   (d) both (b) and (c)
5. The invalid state of an S-R latch occurs when ....
    (a) $S = 1, R = 0$ (b) $S = 0, R = 1$
    (c) $S = 1, R = 1$ (d) $S = 0, R = 0$
6. The purpose of the clock input to a flip-flop is to ....
    (a) clear the device
    (b) set the device
    (c) always cause the output to change states
    (d) cause the output to assume a state dependent on the controlling
         inputs
7. For an edge-triggered D flip-flop ....
    (a) a change in the state of an flip-flop occurs only at a clock pulse
         edge
    (b) the state that the flip-flop goes to depends on the D inputs
    (c) the output follows the input at each clock pulse
    (d) all the above
8. A flip-flop is in the toggle condition when ....
    (a) $J = 1, K = 0$ (b) $J = 1, K = 1$
    (c) $J = 0, K = 0$ (d) $J = 0, K = 1$
9. A stage in a shift register consists of ....
    (a) a flip-flop   (b) a word of storage
    (c) a byte of storage   (d) four bits of storage
10. To serially shift a byte of data into a shift register, there must be ....
    (a) one clock pulse
    (b) one load pulse
    (c) eight clock pulses
    (d) one clock pulse for each 1 in the data
11. To parallel load a byte of data into shift register with a synchronous
    load, there must be ....
    (a) one clock pulse
    (b) one clock pulse for each 1 in the data
    (c) eight clock pulses
    (d) one clock pulse for each 0 in the data

12. The group of bits 10110101 is serially shifted into an 8-bit parallel output shift register with an initial state of 11100100. After two clock pulses, the register contains ....
    (a) 01011110   (b) 10110101
    (c) 01111001   (d) 00101101
13. A ripple counter's speed is limited by the propagation delay of ....
    (a) each flip-flop   (b) all flip-flops and gates
    (c) the flip-flops only with gates   (d) only circuit gates
14. To operate correctly, starting a ring counter requires ....
    (a) clearing all the flip-flops
    (b) presetting one flip-flop and clearing all the others
    (c) clearing one flip-flop and presetting all the others
    (d) presetting all the flip-flops
15. Mod-6 and Mod-12 counters are most commonly used in ....
    (a) frequency counters   (b) multiplexed displays
    (c) digital clocks   (d) power consumption meters
16. Synchronous counters eliminate the delay problems encountered with asynchronous (ripple) counters because the ....
    (a) input clock pulses are applied simultaneously to each stage
    (b) input clock pulses are applied only to the first and last stages
    (c) input clock pulses are applied only to the last stage
    (d) input clock pulses are not used to activate any of the counter stages
17. One of the major drawbacks to the use of asynchronous counters is that ....
    (a) low-frequency applications are limited because of internal propagation delays
    (b) high-frequency applications are limited because of internal propagation delays
    (c) Asynchronous counters do not have major drawbacks and are suitable for use in high- and low-frequency counting applications
    (d) Asynchronous counters do not have propagation delays, which limits their use in high-frequency applications
18. A sequence of equally spaced timing pulses may be easily generated by which type of counter circuit?
    (a) shift register sequencer   (b) clock
    (c) Johnson   (d) binary

19. What happens to the parallel output word in an asynchronous binary down counter whenever a clock pulse occurs?
    (a) The output word decreases by 1
    (b) The output word decreases by 2
    (c) The output word increases by 1
    (d) The output word increases by 2

20. A comparison between ring and johnson counters indicates that ....
    (a) a ring counter has fewer flip-flops but requires more decoding circuitry
    (b) a ring counter has an inverted feedback path
    (c) a johnson counter has more flip-flops but less decoding circuitry
    (d) a johnson counter has an inverted feedback path

21. What type of register would shift a complete binary number in one bit at a time and shift all the stored bits out one bit at a time ?
    (a) PIPO   (b) SISO
    (c) SIPO   (d) PISO

22. Which type of device may be used to interface a parallel data format with external equipment's serial format ?
    (a) key matrix   (b) UART
    (c) memory chip   (d) serial-in, parallel-out

23. When the output of a tri-state shift register is disabled, the output level is placed in a ....
    (a) float state
    (b) low state
    (c) high impedance state
    (d) float state and a high impedance state

24. What is meant by parallel-loading the register?
    (a) Shifting the data in all flip-flops simultaneously
    (b) Loading data in two of the flip-flops
    (c) Loading data in all four flip-flops at the same time
    (d) Momentarily disabling the synchronous SET and RESET inputs

25. What is a shift register that will accept a parallel input and can shift data left or right called ?
    (a) tri-state   (b) end around
    (c) bidirectional universal   (d) conversion

26. What is the difference between combinational logic and sequential logic ?
    (a) Combinational circuits are not triggered by timing pulses, sequential circuits are triggered by timing pulses.
    (b) Combinational and sequential circuits are both triggered by timing pulses.
    (c) Neither circuit is triggered by timing pulses.
    (d) None of these

27. The output of a sequential circuit depends on ....
    (a) present inputs   (b) past input
    (c) both (a) and (b)   (d) none of (a) and (b)

28. The flip-flop is used to store ....
    (a) one bit of information   (b) two bit of information
    (c) one nibble of information   (d) one byte of information

29. The SR flip-flop is also called a ....
    (a) full adder   (b) bistable multivibrator
    (c) astable multivibrator   (d) comparator

30. Which of the following is correct for a gated D flip-flop ?
    (a) The output toggles if one of the inputs is held HIGH.
    (b) Only one of the inputs can be HIGH at a time.
    (c) The output complement follows the input when enabled
    (d) Q output follows the input D when the enable is HIGH.

31. A J-K flip-flop is in a "no change" condition when ....
    (a) J=0 and K=0   (b) J=0 and K=1
    (c) J=1 and K=0   (d) J=1 and K=1

32. Which of the following is Universal flip-flop ?
    (a) J-K flip-flop   (b) R-S flip-flop
    (c) Master slave flip-flop   (d) D flip-flop

33. A D flip-flop can be made from a ....
    (a) JK flip-flop and an inverter   (b) RS flip-flop
    (c) RS flip-flop and an inverter   (d) both (a) and (c)

34. T flip-flop is also called as ....
    (a) Time flip-flop   (b) Transfer data flip-flop
    (c) Toggle flip-flop   (d) None of these

35. Master slave type of flip-flop avoids ....
    (a) designing problem   (b) switching problem
    (c) race around problem   (d) none of these

36. Which of the following flip-flops does not have race around difficulty?
    (a) JK flip-flops   (b) Master slave JK flip-flop
    (c) S-R flip-flop   (d) None of the above

37. Switching time of any flip-flop is a cause of ....
    (a) propagation delay   (b) race around problem
    (c) its functioning   (d) none of these

38. A ripple counter's speed is limited by the propagation delay of ....
    (a) each flip-flop   (b) all flip-flops and gates
    (c) the flip-flops only with gates   (d) only circuit gates

39. How many flip-flop circuits are needed to divide by 8 ?
    (a) Three   (b) Four
    (c) Eight   (d) Sixteen

40. A counter driven by clock can be used to count number of ....
    (a) clock cycles   (b) digits
    (c) propagation delay   (d) none of these

41. A mod 8 counter will count ....
    (a) from 0 to 7   (b) from 0 to 3
    (c) from any number n to n+4   (d) none of above

42. A counter has modulus of 10. The number of flip-flops are ....
    (a) 1   (b) 2   (c) 4   (d) 8

43. In a ripple counter ....
    (a) whenever a flip-flop sets to 1, the next higher FF toggles
    (b) whenever a flip-flop sets to 0, the next higher FF remains unchanged
    (c) whenever a flip-flop sets to 1, the next higher FF faces race condition
    (d) whenever a flip-flop sets to 0, the next higher FF faces race condition

44. The number of flip flops needed for Mod 5 counter are ....
    (a) 7   (b) 5   (c) 1   (d) 3

45. A counter is a ....
    (a) combinational circuit
    (b) sequential circuit
    (c) both combinational and sequential circuit
    (d) none of above

46. How many flip-flops are required to make a MOD-32 binary counter?
    (a) 3   (b) 5   (c) 45   (d) 8

47. The basic shift register operations are ....
    (a) SIPO   (b) PISO
    (c) SISO   (d) All of the above
48. A universal shift register can shift ....
    (a) from right to left
    (b) from left to right
    (c) both from right to left and left to right
    (d) none of above
49. SISO stand for ....
    (a) serial in parallel out   (b) serial in serial out
    (c) series in series out   (d) parallel in serial out
50. Synchronous counter is more faster than Asynchronous counter because ....
    (a) reset pin is high
    (b) common clock applied
    (c) clock pulse applied for only first flip-flop
    (d) none of these
51. PIPO stand for ....
    (a) Parallel in parallel out   (b) Parallel in phase out
    (c) Parallel in series out   (d) None of these
52. The number of states in decade counter is ....
    (a) 4   (b) 8   (c) 10   (d) 16
53. Minimum number of flip-flops required for synchronous decade counter is ....
    (a) 1   (b) 2   (c) 4   (d) 10

Answers

1. (a)	2. (b)	3. (a)	4. (b)	5. (c)	6. (d)	7. (d)	8. (b)
9. (a)	10. (c)	11. (a)	12. (c)	13. (a)	14. (b)	15. (c)	16. (a)
17. (b)	18. (a)	19. (a)	20. (d)	21. (b)	22. (b)	23. (d)	24. (c)
25. (c)	26. (a)	27. (c)	28. (a)	29. (b)	30. (d)	31. (a)	32. (b)
33. (d)	34. (c)	35. (c)	36. (b)	37. (a)	38. (b)	39. (a)	40. (a)
41. (a)	42. (c)	43. (a)	44. (d)	45. (b)	46. (b)	47. (d)	48. (c)
49. (b)	50. (b)	51. (a)	52. (c)	53. (c)			

[B] True or False :

1.   Sequential circuits requires clocks for its operation.

2.   the SR flip-flop is one bit storage cell.

3.   SR flip-flop using NAND gate S = 1, R = 0 is forbidden.

4.   JK FF can be obtained from SR FF.

5.   Preset and clear inputs are used to assign initial states to the FF.

6.   Asynchronous counters are slower than synchronous counter.

7.   Decade counter counts up to 9.

8.   Higher modulus counter.

Answers						
1. True	2. True	3. False	4. True	5. True	6. False	7. True

[C] Short Answer Questions :

1.   What is a flip-flop ? State its use.

2.   What is the limitation of S-R flip-flop ? How it is overcome ?

3.   State the function of preset and clear. Draw the logic diagram of J-K flip-flop with preset and clear.

4.   Show how to convert a J-K flip-flop into D-type. Give its truth table.

5.   Draw the logic diagram of S-R flip-flop converted into D-type. Give its truth table.

6.   Draw the logic diagram to convert J-K into T-type. Give its truth table.

7.   Define the terms positive edge triggered flip-flop and negative edge triggered flip-flop.

8.   Draw the circuit symbol of J-K flip-flop with positive edge triggering and with negative edge triggering.

9.   What is meant by asynchronous counter ? State its limitation.

10. What is meant by synchronous counter ? State its advantages and disadvantages over asynchronous counter.

11. What do you mean by down counter ? Write the count sequence of 3-bit down counter.

12. Define the term modulus of a counter. How many flip-flops will be required to obtain mod 12 counter ?

13. What is a decade counter ? Name the IC which contains the decade counter.

14. State the different applications of the counter.

15. What do you mean by a shift register ?

16. How many flip-flops are required to construct a shift register which can store decimal 100 ?

17. What is the largest decimal and hexadecimal number a 4-bit shift register can store ?

18. State the different applications of shift register.

[D] Long Answer Questions :

1. Explain the working of S-R flip-flop with logic diagram, symbol and truth table.

2. Draw the circuit diagram of J-K flip-flop. Explain its action.

3. Explain what is meant by a race around condition in connection with the J-K flip-flop.

4. Draw a master slave J-K flip-flop system. Explain its operation and show that it eliminates the race around condition.

5. Explain with suitable waveform that one flip-flop does the frequency division by 2.

6. Draw the circuit, truth table and waveforms of 4-bit binary counter.

7. Draw the circuit of 3-bit up/down counter. Explain its action.

8. Explain the function of $R_{0(1)}$, $R_{0(2)}$ and $R_{9(1)}$, $R_{9(2)}$ inputs of IC 7490.

9. Draw the internal block diagram of IC 7490. Explain its action.

10. Draw the circuit diagram of a 3-bit synchronous counter. Explain its operation.

11. Explain the operation of serial in serial out shift register with suitable circuit.

12. Draw the circuit diagram and explain the working of parallel in serial out shift register.

13. What is meant by universal shift register ? State the IC number which contains universal shift register.

14. Explain the operation of ring counter. State its application.

15. Calculate the time delay generated by a 4-bit serial in serial out shift register, if the clock frequency is 1 MHz.

16. Describe how we can use shift register to multiply and divide a binary number.

17. For a 4-bit serial in parallel out shift register, draw a timing diagram to show shifting of data 0111 into it. Assume that register initially contains all 0's.

18. Describe how shift registers are used in time delay generation and in interfacing applications.

19. Draw the circuit of 4-bit bidirectional register. Explain its action.

❑❑❑

MODEL QUESTION PAPERS

MODEL QUESTION PAPER-I

Time : 2 Hrs.   **Marks : 35**

Note :

1. Q. 1 is compulsory.
2. Solve any three questions from Q. 2 to Q. 5.
3. Q. 2 to Q. 5 carry equal marks.

Q.1 Answer the following questions in one or two sentences: (Any 5)

$$(5 \times 1 = 5)$$

(a) Solve for x : $(37)_{10} = (x)_2$.

(b) Simplify using laws of Boolean algebra $y = A\bar{B}C + A\bar{B}\bar{C} + B$.

(c) Write the truth table of half adder.

(d) What is meant by encoder ? Why it is needed.

(e) Write the truth table of decade counter.

(f) State the uses of ring counters.

Q.2 Answer the following :   **(10 Marks)**

(A) What do you mean by universal gates ? Obtain AND gate from only NAND gates.   **(4)**

(B) What is a full adder ? Explain 4-bit parallel adder.   **(6)**

Q.3 Answer the following :   **(10 Marks)**

(A) Explain use of EX-OR gate for parity generation.   **(4)**

(B) What is a decoder ? Explain 3 line to 8 line decoder.   **(6)**

Q.4 Answer the Following :

(A) Explain JK flip-flop with circuit diagram.   **(4)**

(B) What is a shift register ? Explain parallel in serial out shift register.   **(6)**

Q.5 Write a note on : (Any 4)   **(10)**

(a) De Morgan's theorems.

(b) Positive and negative logic.

(c) ASCII code.

(d) K-map for simplification of Boolean equations.

(e) Half adder.

(f) Magnitude comparator.

●●●

(P.1)

MODEL QUESTION PAPER-II

Time : 2 Hrs. **Marks : 35**

Q.1 Answer the following questions in one or two sentences: (Any 5)
$$(5 \times 1 = 5)$$

    (a) Write the truth table of 2 input NAND gate.

    (b) Solve for x : $(72)_{10} = (x)_{16}$.

    (c) What is an half adder ? Write its truth table.

    (d) What do you mean by encoder and decoder ?

    (e) Write the truth table of D-FF.

    (f) State the applications of shift registers.

Q.2 Answer the following : **(10)**

    (A) Explain S-R flip-flop with circuit diagram. **(4)**

    (B) Describe 3-bit asynchronous counter. **(6)**

Q.3 Answer the following : **(10)**

    (A) Explain full adder circuit. **(4)**

    (B) What is meant by multiplexer ? Explain 4 : 1 multiplexer. **(6)**

Q.4 Answer the following : **(10)**

    (A) What do you mean by +ve and −ve logic ? **(4)**

    (B) Explain with truth table and circuit diagram two input NAND gate. **(6)**

Q.5 Write a note on : (Any 4) **(10)**

    (a) Obtain OR gate from only NAND gate.

    (b) Decimal to Binary Conversion.

    (c) Use of K-map.

    (d) De-multiplexer.

    (e) Digital lock using magnitude comparator.

    (f) Ring counter.

•••